Palgrave Studies in Green Criminology

Criminologists have increasingly become involved and interested in environmental issues to the extent that the term Green Criminology is now recognised as a distinct subgenre of criminology. Within this unique area of scholarly activity, researchers consider not just harms to the environment, but also the links between green crimes and other forms of crime, including organised crime's movement into the illegal trade in wildlife or the links between domestic animal abuse and spousal abuse and more serious forms of offending such as serial killing. This series will provide a forum for new works and new ideas in green criminology for both academics and practitioners working in the field, with two primary aims: to provide contemporary theoretical and practice-based analysis of green criminology and environmental issues relating to the development of and enforcement of environmental laws, environmental criminality, policy relating to environmental harms and harms committed against non-human animals and situating environmental harms within the context of wider social harms; and to explore and debate new contemporary issues in green criminology including ecological, environmental and species justice concerns and the better integration of a green criminological approach within mainstream criminal justice. The series will reflect the range and depth of high-quality research and scholarship in this burgeoning area, combining contributions from established scholars wishing to explore new topics and recent entrants who are breaking new ground.

Rachel Killean · Mark Hamilton ·
Kajsa Lundberg
Editors

Conceptualising Justice

A Green Criminological Perspective

palgrave
macmillan

Editors
Rachel Killean
Law School
University of Sydney
Sydney, NSW, Australia

Mark Hamilton
Thomas More Law School
Australian Catholic University
Sydney, NSW, Australia

Kajsa Lundberg
Centre for Urban Research, School of Global, Urban and Social Studies
RMIT University
Melbourne, VIC, Australia

ISSN 2946-269X ISSN 2946-2703 (electronic)
Palgrave Studies in Green Criminology
ISBN 978-3-032-19790-0 ISBN 978-3-032-19791-7 (eBook)
https://doi.org/10.1007/978-3-032-19791-7

This Palgrave Macmillan imprint is published by the registered company Springer Nature Switzerland AG
The registered company address is: Gewerbestrasse 11, 6330 Cham, Switzerland

If disposing of this product, please recycle the paper.

Contents

Notes on Contributors

Dr. Alexander Baird is a Senior Research Fellow at Deakin University. His areas of research interests and expertise include green criminology, environmental crime, and water theft in Australia and transnational environmental crime. He is particularly interested in state and corporate commodification and exploitation of nature and explores the complex and intricate dimensions of environmental harm from a criminological approach. He is currently working on an Australian Research Council Discovery Project entitled: Preventing Water Theft in the Murray Darling Basin.

Dr. Stephen R. Burrell is a Lecturer in Criminology at the University of Melbourne, Australia. His research focuses on men, masculinities and violence—especially the prevention of men's violence against women and the more-than-human world. He is the co-chair of MAEVe, the Melbourne Research Alliance to End Violence against women and their children. He also co-hosts a podcast called 'Now and Men: Current Conversations on Men's Lives', and is an Honourary Fellow at Durham University's Department of Sociology, UK.

Dr. Russil Durrant is a Senior Lecturer at the Institute of Criminology, Victoria University of Wellington in Aotearoa New Zealand. His current research interests focus on green criminology, species justice, and human-invertebrate relations. He is the author of six books, including 'Invertebrate Justice: Extending the Boundaries of Non-Speciesist Green Criminology' (2024, Palgrave).

Dr. Mark Hamilton is a Senior Lecturer in law and criminology within the Thomas More Law School, Faculty of Law and Business, Australian Catholic University. He has taught many different units in both criminology and law including criminal law, environmental law, victimology, restorative justice and green criminology. Mark's research centres around the application of restorative justice to environmental harm, including from sanctioned activity, criminal activity, and activity causing climate change. He has published widely on this topic and has conducted research in both Australia and overseas. His monograph is published in the *Palgrave Studies in Green Criminology* (Restorative Justice and Environmental Crime).

Ms. Erica Honey has been working in the veterinary sector for over 20 years, firstly in private practice as a nurse, then in emergency and critical care for 14 years, leading and managing in a State teaching veterinary hospital. During this time, she also worked in Animal Emergency Management, working in animal emergency planning across sectors, combining her deep knowledge of veterinary and emergency management to create her niche. Erica has also lectured at tertiary institutions on health management, leadership, and people and culture, with a focus on emotional and cultural intelligence, equity, diversity and inclusion. She is a trained and experienced Ally and aims to create safe and inclusive spaces for every person.

Dr. Rachel Killean is a Senior Lecturer at Sydney Law School and the Deputy Director of the Australian Centre for Climate and Environmental Law. Dr Killean's research centres responses to violence, with a focus on transitional justice, victims' rights, sexual and gender-based violence, and harms perpetrated against the natural world. She is the author of Victims, Atrocity and International Criminal Justice (Routledge, 2018); lead editor of Sexual Violence on Trial (Routledge, 2021); and co-author of Green Transitional Justice (Routledge, 2025).

Dr. Kajsa Lundberg is a Research Fellow of Sustainable Consumption and Policy at the RMIT Centre for Urban Research and the TREMS Hub. She holds a Doctoral Degree in Criminology from the University of Melbourne where she has previously held a range of teaching roles, teaching subjects on diverse topics related to cities, political and spatial justice, crime policy, and qualitative research methods. Dr Lundberg's research centres on urban design, environmental harm, waste practices, sustainable consumption, climate mitigation policy, and spatial justice, often applying an interdisciplinary and green criminological approach.

Prof. Amber McKinley is a clinical and forensic victimologist, currently serving as Director of Research in the Centre of Customs and Excise Studies. Her expertise covers victimology, sudden death, human trafficking, and modern slavery, reflected in both her teaching and her collaborations with police, defence, and government agencies. Amber is the creator and director of the National Australian Serial Homicide Registry (NASHR) which holds the largest sample of serial homicide victims, offenders and events in Australia's history, commencing in 1806. Her research informs high-level policy and operational investigations, examining victim vulnerabilities in sexual violence, homicide, and trafficking within Australia and internationally. Her critical homicide research investigates the social, psychological, and situational factors surrounding lethal violence, emphasising victim-offender relationships, socio-economic contexts, and behavioural patterns. By examining patterns of victimisation and the strategies employed, her research informs targeted prevention strategies and enhances solvability through victim-centred approaches.

Dr. Cassie A. Pedersen is a Lecturer in Criminology and Criminal Justice at Federation University, Australia. Her work centres on exposing extraordinary harms embedded in ordinary and routine practices. She approaches these issues through the lenses of trauma theory, green criminology, ecofeminist theory, and critical animal studies. Within these interdisciplinary frameworks, she examines how systemic violence against the more-than-human world is normalised and obscured in everyday contexts.

Dr. Michael Scott Smith is an early-career researcher from Griffith University in Australia and an adjunct member of Griffith Criminology Institute. Dedicated to wildlife conservation, he is interested in media criminology, environmental enforcement, green criminology and environmental communication regarding the rhino horn trade in Vietnam. With 15 years' experience investigating the rhino horn trade, he holds a Ph.D., a Master of Arts Research (MAR) and a Master of Journalism. Prior to his doctoral research into the news framing of the rhino horn trade, Michael worked in Vietnam as a journalist and sub-editor for national and metro newspapers. He investigated efforts to protect the now extinct Javan Rhino population in Vietnam, the demand for rhino horn in Ho Chi Minh City and then investigated Vietnamese illegal syndicates' role in rhino poaching in South Africa. His broader research interests branch into the illegal wildlife trade worldwide.

Dr. Astrid Vachette is a lecturer and discipline lead in Emergency Management at the Australian Graduate School in Policing and Security, Charles Sturt University. She holds a Bachelor and Masters from France in international disaster management and completed her doctoral thesis at the Centre for Disaster Studies, James Cook University, on integrated disaster resilience in Vanuatu. She also optimised a COVID-gap-year and completed a graduate diploma in Psychology at the University of New England, to better integrate mental health considerations into her teaching and research. Astrid is a volunteer with New South Wales Rural Fire Services and trained crisis supporter. She is an interdisciplinary social scientist exploring the drivers and challenges of resilience within the social ecology of human development. Her current research includes: sustainability in disaster management, community organising, networks of climate and disaster knowledge systems, relational climate justice, climate victimology, and Nature-based Solutions for floods.

Prof. Reece Walters joined Deakin University in November 2018. He has previously held academic appointments at QUT, The Open University, University of Stirling, Victoria University of Wellington and La Trobe University. His areas of research interests and expertise include Green Criminology, Crimes of the Powerful, Environmental Law and the Sociology of Knowledge. He is particularly interested in the ways in which corporations and governments exploit and compromise the 'essentials of life', namely air, food and water for power and profit.

Dr. Sarah Wright is Senior Lecturer with the School of Law at the University of Wollongong, Australia. Sarah's research focus includes environmental crime, compliance and enforcement in environmental law, regulatory theory and effective regulation. Prior to entering academia Sarah worked as a Senior Legal Officer with the NSW Office of Environment and Heritage, which then incorporated the NSW Environment Protection Authority, and also as Tipstaff to the Hon. Justice Nicola Pain of the Land and Environment Court of NSW.

Prof. Rob White is Emeritus Distinguished Professor of Criminology at the University of Tasmania and Professor of Criminology at the University of New England, Australia. He has written extensively in criminology, youth studies and eco-justice. Recent books include Theorising Green Criminology (Routledge, 2022), Advanced Introduction to Applied Green Criminology (Edward Elgar, 2023) and the Elgar Encyclopedia of Environmental Crime (editor, Edward Elgar, 2025).

List of Tables

1

Expanding Justice In and Through Green Criminology

Rachel Killean, Kajsa Lundberg, and Mark Hamilton

1 Introduction

This book critically assesses and furthers the debates surrounding the meaning of 'justice' in a world grappling with intersecting ecological crises. In response to the growing urgency of the interconnected climate, biodiversity, and pollution crises, scholars across the globe are rethinking how to conceptualise 'justice' in relation to human and more-than-human victimhood, harm, repair, and resistance. Within the framework of green criminology, this reimagining often entails a shift away from

R. Killean
University of Sydney, Sydney, NSW, Australia
e-mail: rachel.killean@sydney.edu.au

K. Lundberg (✉)
RMIT University, Melbourne, VIC, Australia
e-mail: kajsa.olivia.lundberg@rmit.edu.au

M. Hamilton
Australian Catholic University, Sydney, NSW, Australia
e-mail: mark.hamilton@acu.edu.au

R. Killean et al. (eds.), *Conceptualising Justice*, Palgrave Studies in Green Criminology, https://doi.org/10.1007/978-3-032-19791-7_1

paradigms centred on humans and crime. As this book illustrates, green criminology leverages its interdisciplinary roots to challenge traditional understandings of justice, reshaping both theory and practice.

A wide and expanding array of justice frameworks has emerged in the field, including restorative justice, environmental justice, climate justice, multi-species justice, reparative justice, ecological justice, water justice, and even invertebrate justice. Important contributions to these debates have originated from the Antipodes (Australia and New Zealand/ Aotearoa), a region profoundly affected by compounding environmental challenges. Emerging from discussions within the Australia and New Zealand Society of Criminology's Green Criminology group (which is convened by the editors), this book brings together established and emerging scholars from across New Zealand/Aotearoa and Australia to examine these diverse and evolving notions of justice through the lens of green criminology. In doing so, it explores how these interconnected approaches might align, overlap, or inform one another to advance the cause of more-than-human justice. Across its justice-driven chapters, the book interrogates the implications of these frameworks for the future of green criminology as a discipline and critically examines how these justice-oriented perspectives may challenge the very concept of 'criminology'.

Outside the context of academic scholarship, legal and ethical challenges further highlight the necessity of rethinking and interrogating notions of 'justice'. Current legal frameworks often fail to adequately address environmental harm or hold powerful corporations and states accountable for ecological destruction. At the same time, balancing economic development with environmental conservation raises profound ethical questions. What obligations do we owe to non-human entities, future generations, and communities already bearing the costs of environmental harm? How can justice systems evolve to reflect these obligations?

This book contributes to bridging the gap between theory and practice, bringing together a range of perspectives to offer practical insights for policymaking, legal reform, and grassroots activism. By addressing the philosophical, legal, and sociopolitical dimensions of justice, we hope that this collection may usefully inform movements seeking to combat environmental and ecological injustices. In an era defined by ecological crisis, we believe that the concept of justice requires ongoing interrogation to advance our ability to meet the challenges of the Anthropocene.

This introductory chapter proceeds as follows. In Part 2, we begin by tracing the origins and evolution of green criminology as a distinct area of scholarship. As we explore, the emergence of green criminology in the 1990s is rooted in critical theoretical debates about how 'crime', 'harm' and 'justice' are defined and whether and how ecological harm fits within these constructs. As we further note, it can also be contextualised in the growing awareness across public and policy discourse regarding the urgency of addressing environmental harms. The chapter highlights the field's evolution into a critical, interdisciplinary, and dynamic domain, with scholarship emerging from around the world which challenges and expands established understandings of concepts such as crime, perpetration, victimhood, harm, and repair (Hall, 2014; Long et al., 2024; White, 2018).

In Part 3, the chapter introduces the diversity of justice frameworks that have emerged within green criminology. It highlights how green criminology has offered a sustained critique of and an alternative to dominant anthropocentric and retributive notions of justice. We explore how green criminology challenges the human-centred nature of dominant justice theories, and how scholars have highlighted the failure of dominant theories to account for the full spectrum of harm caused by environmental harm, particularly when it affects non-human entities. We provide some illustrative examples of the range of justice formulations that have been put forward within green criminology, encompassing environmental justice, species justice, ecological justice, interspecies justice, and climate justice, highlighting their grounding in different understandings of harm and responsibility. As we demonstrate, these conceptualisations offer varied accounts of who or what deserves justice and how justice might be meaningfully pursued. While some approaches remain rooted in criminal law, others advocate for a shift towards healing and prevention rather than punishment. Together, these threads of scholarship illustrate the trajectory of green criminology towards a more diverse, inclusive, and critical understanding of justice in the context of environmental harm.

This chapter concludes by introducing the subsequent chapters, which build upon these foundations to delve deeper into specific issues, methodologies, and case studies primarily from Australia and New Zealand, offering a rich and multidisciplinary perspective on justice within green criminology.

2 The Evolution of Green Criminology

Before taking a closer look at green criminology's evolving relationship to questions of justice, it is worth taking some time to sketch out the field itself. References to 'green criminology' first emerged in English-language scholarship in the early 1990s, as several criminologists began to question why the field had not traditionally seen environmental harms as falling within criminology's field of enquiry. By that time, 'green' issues had emerged in various other fields of study across the social sciences and humanities, as well as in broader political and public consciousness (political economy was a particular influence, see Lynch, 1990, 2020). Although not explicitly framed as 'green criminology', criminological studies had investigated pollution, corporate crime, organised crime, military impacts on the environment, and the policing of harms to the natural world (South, 2014: 6). As Goyes and South (2017) have explored, relevant research was also taking place in non-English scholarship, including in Scandinavian states (Christophersn & Johansen, 1992), Slovenia (Pečar, 1981), Switzerland (André, 1980), and Spain (Rodríguez Ramos, 1981). Early English-language interventions were therefore reflective of broader intellectual and societal currents of the time (South, 2014: 8), although they did not always connect to the research taking place across other disciplines and contexts.

Thus, the 1990s brought not the first studies in these areas, but the first explicit arguments that crimes against the environment were of global importance, that they were worthy of more attention and rigorous theorisation, *and* that they might constitute their own specific field of study within the criminology field. The goal of these initial interventions was often theoretically modest. As South (2014) explains, the aim was rarely to put forward a definitive 'theory' of green criminology, although various theoretical orientations emerged (Lynch & Stretesky, 2003). Rather, the argument was often more simply that criminologists should take greater notice of environmental concerns. At the same time, early green criminology scholarship was ambitious in that it both invited greater critical reflection on what counts as 'crime' (Lynch, 1990) and called for interdisciplinary work that drew from social and natural sciences (Beirne & South, 1998). From the start, green criminologists were aligned with more critical approaches (Koser Wilson, 1999), drawing from deviancy theories, Marxist criminology, feminist criminology and critical criminology to emphasise crimes of the powerful and the plight of the marginalised and voiceless (South, 2014: 7–8).

Over the subsequent decades, green criminology has become more established, as evidenced by this book series and the Routledge Green Criminology book series, specialist groups around the world (such as ANZSOC and the Green Criminology European Working Group), bespoke conferences and thousands of publications. Green criminology continues to avoid a definitive theoretical position, although it remains predominantly grounded in critical perspectives (Sollund, 2021). It might rather be seen as an umbrella term or framework (Brisman & South, 2013), bringing together a diversity of perspectives (including from multiple disciplines) to engage with crime, harm and injustice related to the environment and other-than-human species (Lynch & Stretesky, 2014). Within this framework, numerous sub-fields or focuses of enquiry have emerged, including some discussed further in this book (such as White's eco-criminology, discussed in Chapter Two). Green criminology is taking place around the world, with scholars focusing both on global challenges such as the climate crisis and on locally specific problems (Sollund, 2023). Sub-fields such as 'Southern green criminology' (Goyes, 2016), for example, draw explicit attention to how environmental harms play out in the global south.

The field continues to apply established criminological questions and approaches to emerging contexts. Much of green criminological scholarship engages with questions regarding the distinctions between harm, civil violation, and crime (Sollund, 2021). Such questions regarding the social construction of crime are a longstanding feature of criminological scholarship (Quinney, 1970) but are especially salient in the context of environmental destruction, which is frequently carried out within legal frameworks (Lynch & Stretesky, 2014; Wyatt, 2013). While emphasising power imbalances, and state-corporate pollution, this literature also highlights the often-custodial nature of environmental harm which comes to blur boundaries between victims and offenders (Agnew, 2013; Lundberg, 2022). Similarly, pre-existing questions as to who counts as a legitimate 'victim' of crime and what that recognition entails (Christie, 1986; Weis & Borges, 1973) have been fruitfully extended to consider questions of speciesism and the exclusion of other-than-human victims from view (Hall, 2014; White, 2018).

The application of existing theories and questions should not be taken to mean that the emergence of green criminology is not profoundly radical in its implications. By drawing on existing concepts and interrogating their application to the natural world, criminologists have stretched the boundaries of criminological thinking, broadening its focus to 'green' questions around how harm is constructed, who is held accountable, and how justice can be conceptualised and delivered in cases of environmental degradation. The following section picks up this last theme, exploring how green criminologists have interrogated 'justice' as a concept, critiqued existing justice processes, and sought to diversify what justice entails.

3 Green Criminology and Justice

Green criminology theories of justice are often offered as a critique of traditional notions of criminal justice. A particular longstanding critique relates to the anthropocentrism of mainstream criminal justice (Halsey & White, 1998). As argued by Lynch et al. (2019: 127–128), 'theories of justice that appear within the orthodox criminological and criminal justice literature are always anthropocentrically derived or constructed entirely in relation to human social relationships, values and expectations'. Such approaches, green criminologists often argue, neglect ecological systems, and non-human entities, resulting in incomplete responses to environmental harm and non-human suffering. There are different threads of this critique evident in the literature, illustrating scholars' focus on differing aspects of harm. Scholarship highlights the failure to account for the human harm experiences as a result of environmental damage (Lynch et al., 2015; Pellow, 2004), harm against animals (e.g. Beirne, 2009) and against specific natural systems, as well as harms perpetrated against earth as a whole living system (Lynch et al., 2019). As Benton (2007) has articulated, recognising these harms and their implications poses substantial challenges for established thinking about justice, raising fundamental questions regarding what humans might owe to non-human entities and to one another, as well as how different (and potentially competing) justice claims are to be addressed (see also White, 2016a).

Lynch et al. (2019: 129) note that an anthropocentric framing omits 'alternative possibilities for grounding theories of justice in more universalistic terms'. Green criminology has been generative as well as critical, offering alternative ways of thinking about justice, conceptually and practically. Mapping onto the areas of concern noted above, green articulations of 'justice' have varied depending on theorists' framing of the problem. For example, some green criminological literature is grounded in 'environmental justice', a framing of justice primarily concerned with pursuing human communities' equal access to environmental benefits and ensuring equal protection from environmental ills (Bullard, 1990). Often retaining an anthropocentric orientation, environmental justice draws attention to the ways in which particular communities, primarily those who are already marginalised, are discriminately exposed to pollution and other environmental burdens (Mohai et al., 2009). A more recent articulation of this concern is found in the concept of 'climate justice' which highlights the unequal impacts of climate change (Brisman et al., 2018; White & Kramer, 2015).

Looking beyond humans, 'species justice' instead centres harms perpetrated against non-human species (usually but not exclusively animals), highlighting their inherent value, arguing for their procedural and substantive rights, and drawing attention to the multiple ways in which non-human animals might be reframed as victims of crime (Lie &

Sollund, 2017). Enriched by parallel developments in animal rights and rights of nature scholarship, concerns with non-human justice have more recently evolved to encompass notions of 'multi-species' or 'interspecies' justice, conceptualisations that consider a range of justice claims and the relationships between them, rather than the claims of a specific non-human species (Celermajer et al., 2025). An overlapping conceptualisation of justice has been that of 'ecological justice', an approach which similarly recognises the interconnectedness of human and non-human species and demands that human action be 'evaluated in relation to potential harms and risks to specific creatures and specific locales as well as the biosphere generally' (White, 2008: 19).

Rob White has played a central role in articulating the above justice approaches. In his influential 2013 book, he interrogates the emergence and development of environmental justice, species justice and ecological justice, drawing them together under the umbrella concept of 'eco-justice' (White, 2013). He has continued to use this concept, which he explores in Chapter Two, to consider the moral principles that might guide a more inclusive justice, recognising the conflicting interests and contradictory situations that a more inclusive approach entails (White, 2016a). As he unpacks in his chapter, new environmental challenges raise the possibility of new justice formulations, and this has indeed been the case in the green criminology scholarship. New justice formulations emerge all the time in response to ecological crises, with notable examples including 'astro-green criminology', which seeks to explore questions of justice and injustice in the context of outer space (Eski & Lampkin, 2024).

Reflecting its critical orientation, green criminology offers a fundamental critique of the limits of criminal justice. Building on its analysis of how 'crime', 'harm', 'victimhood', and 'perpetration' are defined, green criminological scholarship frequently highlights how criminal justice is socially constructed to preclude justice for 'lawful but awful' acts perpetrated against the natural world (Hamilton, 2022). Green criminology is not without a retributive angle; there is scholarship that explicitly seeks to expand criminal law to encompass a wider range of environmental harms, including through an international crime of ecocide (Killean,

2025; Short, 2016; White & Kramer, 2015). There is also some literature on how to 'green' criminal justice mechanisms, such as through the phenomenon of the 'green prison' (critiqued by Mazurek et al., 2020). However, it is also notable that scholars have sought to move beyond a purely retributive approach to the perpetration of environmental harm (de Nardin Budó et al., 2023). In doing so, green criminologists again turn to concepts that appear elsewhere in criminological thought, while evolving and expanding their meaning to encompass 'green' issues. This manifests both in engagement with whether and how criminal law should respond to acts of environmental harm (Brisman & South, 2013; White, 2008), and in scholarship that rejects a focus on crime entirely and argues for a reorientation around harm and deviance (e.g. Lynch & Stretesky, 2003).

The willingness to look beyond punishment and to expand notions of justice has implications for justice's processes and outcomes. With regard to process, 'green victimologists', amongst others, have argued for justice processes to better account for the needs of non-human entities, including through recognition of their victimhood and associated access to victims' rights (Flynn & Hall, 2017; Hall, 2014; White, 2018). Such arguments find resonance with rights of nature scholarship, with both bodies of literature interrogating practical but also deeply fundamental questions concerning how to best represent non-human interests in a justice setting (Killean & Newton, 2025). In terms of outcome, an expansion of justice beyond the human similarly raises challenging questions regarding how to determine and deliver appropriate outcomes (White, 2016b). In seeking solutions beyond a traditional focus on punishment, green criminologists often return to the possibilities of more reparative (Killean, 2021; White, 2017) and restorative justice (Hamilton, 2021; Natali & Hall, 2021; Varona, 2021) approaches. While distinct in some ways (restorative justice requiring the consent of all participants, reparative justice being something that can be pursued without the perpetrators' consent), these approaches both reflect a shift towards ecological healing, restoration, and relationship-building, evidencing green criminologists' concern with how to pursue ecological restoration, transform relationships and prevent future harm.

Taken together (and this overview only scratches the surface of the rich literature in this space), this ever-growing scholarship illustrates how conceptualisations of justice are proliferating in response to the complex realities of environmental harm. Emerging as a critique of anthropocentrism and traditional criminal justice, green criminological literature has evolved to offer a rich array of justice imaginaries. This book contributes to this expansive scholarship, identifying, interrogating, and expanding on notions of justice.

4 Conceptualising Justice

The rest of the book digs deeper into the three themes identified above: problematising justice, justice as inclusion and participation, and just outcomes. As will be seen, many of the concerns and approaches to justice explored in the preceding section are picked up and explored in more depth by our authors.

The book proceeds as follows. The first three chapters problematise the meaning of 'justice'. In Chapter Two, Rob White provides a detailed exposition of the key elements of eco-justice, a concept that White has developed through extensive research. He begins by discussing the origins of the eco-justice framework and its evolution from inception to the present day. White's original eco-justice framework was comprised of three key dimensions: environmental justice (humans), ecological justice (ecosystems), and species justice (non-human animals). This emerged out of his analysis of diverse environmental and animal rights movements. As he notes, each type of social movement has tended to focus on specific categories of victims. Together, they cover humans as well as non-human environmental entities such as animals, rivers, and trees. White argues that this inclusiveness means that research attention is directed to matters of habitat, species, and landscape, as well as human communities. White unpacks how, over time, his eco-justice framework has expanded to incorporate plants, and cosmological entities (for example, the Moon), into its broad perspective. The implications of this inclusion are discussed in relation to biodiversity, climate change and extraterrestrial bodies, as well as animal welfare and rights. White concludes by noting the necessity for and complexity of weighing up diverse rights and interests in concrete situations.

In Chapter Three, Alexander Baird and Reece Walters unpack the concept of 'water justice'. Baird and Walters define water justice as the fair and equitable distribution, access, and management of water resources, with a focus on addressing social, environmental, and economic inequalities related to water. It is based on the principle that water is a fundamental human right and should be accessible to everyone, regardless of social status, income, or geographic location. As Baird and Walters explore, water justice is often intertwined with larger social justice movements, including poverty reduction, climate justice, and Indigenous rights, and it requires the intersection of legal, social, environmental, and economic discourses and solutions. The chapter conceptualises an understanding of water justice through the lens of green criminology, drawing from eco-philosophical frameworks of environmental, ecological and species justice to critically explore and compare what water justice means and its associated issues within the Australian context, namely the Murray-Darling Basin (the country's largest and most regulated river system) as well as coastal estuaries and marine ecosystems.

In Chapter Four, Cassie Pedersen and Stephen Burrell use ecofeminism to interrogate gendered approaches to justice for non-human animals and the environment. They argue that an ecofeminist framework identifies and juxtaposes two gendered approaches: a masculinised 'justice' approach and a feminised 'care' approach. Advocating for an ecofeminist approach, Pedersen and Burrell challenge abstract, detached, and masculinised notions of 'justice'. Instead, they promote feminist, relational, and contextual ontologies through the concept of 'care'. Their approach does not seek to discard the concept of 'justice' altogether, nor does it uphold an either/or logic in which one must choose between 'justice' or 'care'. Instead, they formulate a concept of 'just care', a concept that involves using one's particular embodied experience as the basis for drawing more general, abstract inferences. It entails caring for and about concrete others and using this as a foundation for the broader pursuit of justice—including for the more-than-human world. Pedersen and Burrell apply this approach to Tasmanian salmon farming, illustrating how an ecofeminist lens can make sense of the multifaceted harms underpinning this industry and guide green criminological responses.

Chapter Five, by Sarah Wright, turns from problematising justice to questions of inclusion and participation. Her chapter starts from the premise that participation by way of humans speaking for the environment and non-human species is an important component of eco-justice in practice. Wright considers the role that the public has and could play in protecting and speaking for the environment in criminal prosecutions by providing evidence and through restorative justice conferencing. Using New South Wales, Australia as a case study, Wright's chapter examines the role and limitations of mechanisms by which the public can speak for the environment in criminal prosecutions. She considers existing mechanisms, namely acting as prosecutor, giving evidence as a witness or victim, and restorative justice conferencing. Her chapter then examines a recent prosecution where an environmental group was permitted to speak for koalas and the environment in a sentencing hearing as amicus curiae: 'a friend of the court'. Wright reflects on the potential benefits of this approach in operationalising eco-justice, but also concerns regarding impact on the operation of the criminal justice system, including trespass into the prosecutor's discretion in presenting the case and fairness to the defendant.

In Chapter Six, Mark Hamilton explores the concept of justice as meaningful involvement. In the context of environmental offending, justice as meaningful involvement can be achieved through the use of restorative justice conferencing, a face-to-face dialectic exchange. Given green criminology's focus on illegal activity as well as legal activity, Hamilton notes the temptation to simply transplant restorative justice conferencing as used in an environmental offending context for use with environmentally harmful legal activity. If such a use is proposed, it is important to recognise that justice as meaningful involvement means different things in the context of environmental offending and in the context of lawful activity which harms the environment. Hamilton's chapter frames justice as meaningful involvement in the context of environmentally harmful but legal activity. Such a framing suggests that restorative justice conferencing faces challenges in trying to achieve justice in the context of legal activity which harms the environment.

In Chapter Seven, Amber McKinley, Astrid Vachette, and Erica Honey explore the growing legal recognition of animal sentience, that is, the capability of subjective perception, which in turn means that animals are deserving of compassion. Their chapter explores the legal recognition of animal sentience in the Australian Capital Territory's *Animal Welfare Legislation Amendment Act 2019* (ACT) specifically and like legislation internationally, using green criminology's eco-justice framework. They investigate how the ACT Act's explicit acknowledgement of animals as sentient beings challenges the traditional legal view of animals as 'property', as seen in statutes like the *Crimes Act 1900* (ACT). The analysis draws on green criminological theory to situate this reform within debates about species justice, which argue for treating harmed non-human animals as victims rather than mere property. McKinley, Vachette and Honey discuss the Act's progressive stance on animal mental welfare and intrinsic value, while also highlighting ongoing tensions with anthropocentric legal frameworks. They assess the Act's practical impacts, such as expanded duty-of-care obligations and bans on prolonged confinement, but critiques industry exemptions under codes of conduct. By placing these reforms within broader ecological justice debates, their chapter emphasises both the symbolic and incremental importance of recognising animal sentience in law, while warning against regulatory approaches that maintain human dominance over animals.

In Chapter Eight, Kajsa Lundberg employs a spatial justice approach to consider how green criminological concerns are at play in and around the flammable and floodable city. As she explores, spatial justice, building on work on the right to the city and the social production of space, brings together justice/injustice concerns with spatial and geographical elements such as access to safe and liveable spaces, environments, mobility, and exclusion/inclusion. Lundberg highlights the applicability of the concept to work seeking more equitable cities, including concerns around affordable housing, public transport systems, and urban population patterns. Drawing from insights, policies, and reports associated with the 2019/2020 Black Summer fires and recent flooding events in Australia, Lundberg sketches out the spatial injustices associated with urban political and planning responses, drawing attention to how such concerns overlap with green criminological work on more-than-human harm and extreme weather events. In emphasising the importance of recognising the spatial inequalities associated with urban development, Lundberg argues for the importance of addressing and mitigating the harm caused by development and housing in Australian cities amidst a changing (urban) climate.

Chapter Nine, by Russil Durrant, draws on non-speciesist green criminology to argue for the importance of invertebrate justice. Durrant presents an Invertebrate Justice Model that outlines humans' direct and indirect justice obligations towards invertebrates. The chapter focuses on the importance of animal welfare law in better realising justice outcomes for invertebrates. In recent years, there has been a shift in animal law to better recognise the importance of sentience in making decisions about which animals should be included. There is also a burgeoning literature on the capacity for sentience in many insect species. Durrant argues that these developments open up the opportunity to bring insects within the ambit of animal welfare law. To explore this possibility, he argues for the inclusion of insects in the *Animal Welfare Act 1999* (NZ), and what this might mean to insects themselves, humans, and the wider environment.

Chapter Ten shifts from participation and inclusion to the consideration of just outcomes. In this chapter Michael Smith explores whether and how the introduction of jail sentences of arrested rhino horn trade kingpins and couriers from 2016 to 2020 in Vietnam provided justice for rhinos in Africa that were killed for their horns. The chapter uses green cultural criminology and news framing to explore the narratives used to portray the Vietnamese rhino horn trade. It uses press releases sent to Vietnamese journalists from environmental Non-Governmental Organisations and articles in Vietnamese digital newspapers. Smith's findings demonstrate the impact that international stakeholder and NGO pressure through local Vietnamese media has possibly had on enforcement of laws against the rhino horn trade in Vietnam. Through persistent news framing, international stakeholders, and NGOs have highlighted the lack of compliance in Vietnam and encouraged the strengthening of legal frameworks to combat the rhino horn trade. From a criminality-focused understanding of justice, which centres focuses on the notion of deterrence through tougher crime-based approaches to harm, Smith argues that these pressures have provided some justice for rhinos in Africa.

In Chapter Eleven, Rachel Killean explores the emergence of a global campaign to criminalise ecocide as an international and domestic crime. The chapter asks two questions. First, what visions of 'justice for ecocide' are campaigners seeking when they pursue criminalisation, i.e. what outcomes are they seeking? Second, is the criminalisation of ecocide equipped to deliver these outcomes in practice? The chapter draws on publicly available material to consider the purported justice goals of those who champion criminalisation, finding a strong emphasis on deterrence and retribution. To answer the second question, Killean analyses and problematises the turn towards criminalisation within human and environmental rights, reflecting on the limitations of criminalisation as a means of addressing and redressing the destruction of the natural world. In particular, it critiques the neoliberal emphasis on the 'individual', noting the multifaceted and interconnected nature of ecocide perpetration and victimisation. Killean argues that practical and structural challenges are likely to place tight parameters around the ability of a new crime to deliver on its promises of justice.

In the final chapter, we (the editors) draw together the themes that emerged over the course of the book. To do so, we draw on the concept of Kaleidoscopic Justice, developed by Clare McGlynn and Nicole Westmarland (2019) in the context of sexual violence. McGlynn and Westmarland use this framework to challenge traditional understandings of justice as something linear, generally retributive, and often delivered by state structures. As they highlight, 'justice' is better understood as 'a continually shifting pattern' and 'a lived, ongoing and ever-evolving experience and process'. Applying this framework to green crime, harm, and more-than-human justice concerns, we argue, allows us to think through what a more pluralistic and adaptive response to environmental destruction might entail. Moreover, drawing from Science and Technology Studies, we highlight the importance of shifting attention from anthropocentric conceptions of agency as something inherently human (with particular justice implications) to consider the ways humans, non-human species, ecologies, and environments co-constitute the world and hierarchies within it. Reflecting the themes explored throughout the book, we sketch out an overlapping and interlinked approach that prioritises inclusivity, the participation of victims (human and non-human), dignity, meaningful consequences, and—importantly—relationality.

References

Agnew, R. (2013). The Ordinary Acts that Contribute to Ecocide. In N. South & A. Brisman (Eds.), *Routledge International Handbook of Green Criminology* (pp. 58–72). Routledge. https://doi.org/10.4324/9780203093658.ch3

André, J. D. (1980). Délinquance Écologique: De l'Artificiel au Transactionnel. *Déviance Et Société, 4*(4), 399–412.

Beirne, P. (2009). *Confronting Animal Abuse*. Rowman & Littlefield.

Beirne, P., & South, N. (Eds.) (1998). For a Green Criminology, Special Issue of *Theoretical Criminology, 2*(2).

Benton, T. (2007). Ecology, Community and Justice: The Meaning of Green. In P. Beirne & N. South (Eds.), *Issues in Green Criminology* (pp. 3–32). Routledge.

Brisman, A., & South, N. (2013). Introduction: Horizons, Issues and Relationships in Green Criminology. In N. South & A. Brisman (Eds.), *Routledge International Handbook of Green Criminology* (pp. 1–24). Routledge.

Brisman, A., South, N., & Walters, R. (2018). Southernizing Green Criminology: Human Dislocation, Environmental Injustice and Climate Apartheid. *Justice, Power and Resistance, 2*(1), 1–21. https://doi.org/10536/DRO/DU:30116302

Bullard, R. D. (1990). *Dumping in Dixie: Race, Class, and Environmental Quality*. Westview Press.

Celermajer, D., Burke, A., Fishel, S., Fitz-Henry, E., Rogers, N., Schlosberg, D., & Winter, C. (2025). *Institutionalising Multispecies Justice*. Cambridge University Press.

Christie, N. (1986). The Ideal Victim. In E. Fattah (Ed.), *From Crime Policy to Victim Policy* (pp. 17–30). Palgrave MacMillan.

Christophersen, J. G., & Johansen, P. O. (1992). Forurensing som Tradisjon. In P. O. Johansen (Ed.), *Studier i Økonomisk Kriminalitet* (pp. 254–291). Institutt for Kriminologi.

de Nardin Budó, M., França, K., & Natali, L. (2023). Beyond Retributive Justice: Listening to Environmental Victims' Demands in Brazil. In D. R. Goyes (Ed.), *Green Crime in the Global South* (pp. 211–241). Palgrave.

Eski, Y., & Lampkin, J. (2024). *Crime, Criminal Justice and Ethics in Outer Space: International Perspectives.* Routledge.

Flynn, M., & Hall, M. (2017). The Case for a Victimology of Nonhuman Animal Harms. *Contemporary Justice Review, 20*(3), 299–318. https://doi.org/10.1080/10282580.2017.1348898

Goyes, D. R. (2016). Green Activist Criminology and the Epistemologies of the South. *Critical Criminology, 24*(4), 503–518. https://doi.org/10.1007/s10612-016-9330-y

Goyes, D. R., & South, N. (2017). Green Criminology Before 'Green Criminology': Amnesia and Absences. *Critical Criminology, 25*, 165–181. https://doi.org/10.1007/s10612-017-9357-8

Hall, M. (2014). Environmental Harm and Environmental Victims: Scoping out a 'Green Victimology.' *International Review of Victimology, 20*(1), 129–143. https://doi.org/10.1177/0269758013508682

Halsey, M., & White, R. (1998). Crime, Ecophilosophy and Environmental Harm. *Theoretical Criminology, 2*(3), 345–371. https://doi.org/10.1177/1362480698002003003

Hamilton, M. (2021). *Environmental Crime and Restorative Justice: Justice as Meaningful Involvement*. Palgrave. https://doi.org/10.1007/978-3-030-69052-6

Hamilton, M. (2022). Restorative Justice Conferencing: A Vehicle for Repairing Harm Emanating from Lawful but Awful Activity. In J. Gacek & R. Jochelson (Eds.), *Green Criminology and the Law* (pp. 361–386). Palgrave. https://doi.org/10.1007/978-3-030-82412-9_16

Killean, R. (2021). From Ecocide to Eco-Sensitivity: 'Greening' Reparations at the International Criminal Court. *The International Journal of Human Rights, 25*(2), 323–347. https://doi.org/10.1080/13642987.2020.1783531

Killean, R. (2025). Ecocide's Evolving Relationship with War. *Environment and Security*. https://doi.org/10.1177/27538796251347111

Killean, R., & Newton, E. (2025). From Ecocide to Ecocentrism: Conceptualising Environmental Victimhood at the International Criminal Court. *International Review of Victimology, 31*(2), 238–264. https://doi.org/10.1177/02697580241269426

Koser Wilson, N. (1999). Eco-Critical Criminology—An Introduction. *Criminal Justice Policy Review, 10*(2), 155–160. https://doi.org/10.1177/0887403499010002002

Lie, M. S. B., & Sollund, R. (2017). Speciesism and Theriocide. In A. Brisman, E. Carrabine, & N. South (Eds.), *The Routledge Companion to Criminological Theory and Concepts* (pp. 386–390). Routledge.

Long, M. A., Stretesky, P. B., & Lynch, M. J. (2024). Green Criminology. In D. A. Gill, L. A. Ritchie., & N. M. Campbell (Eds.), *Encyclopaedia of Technological Hazards and Disasters in the Social Sciences*. Elgar.

Lundberg, K. (2022). Walking at the Edges of Green Criminology: The Edges of the City and the Extraordinary Consequences of Ordinary Harms. *Criminological Encounters, 5*(1), 18–34. https://doi.org/10.26395/CE22050103

Lynch, M. J. (1990). The Greening of Criminology: A Perspective on the 1990s. *The Critical Criminologist, 2*(3), 1–12.

Lynch, M. J. (2020). Green Criminology and Environmental Crime: Criminology that Matters in the Age of Global Ecological Collapse. *Journal of White Collar and Corporate Crime, 1*(1), 50–61. https://doi.org/10.1177/2631309X19876930

Lynch, M. J., & Stretesky, P. B. (2003). The Meaning of Green: Contrasting Criminological Perspectives. *Theoretical Criminology, 7*(2), 217–238. https://doi.org/10.1177/13624806030072414

Lynch, M. J., & Stretesky, P. B. (2014). *Exploring Green Criminology: Toward a Green Criminological Revolution*. Ashgate Publishing, Ltd.

Lynch, M. J., Long, M. A., & Stretesky, P. B. (2019). Gaia and a Green Theory of Justice. In M. J. Lynch, M. A. Long, & P. B. Stretesky (Eds.), *Green Criminology and Green Theories of Justice* (pp. 127–149). Palgrave Macmillan.

Lynch, M. J., Stretesky, P. B., & Long, M. A. (2015). Environmental Justice: A Criminological Perspective. *Environmental Research Letters, 10*(8), article 085008. https://doi.org/10.1088/1748-9326/10/8/085008

Mazurek, J. E., Piché, J., & Schept, J. (2020). 'Greening' Injustice: Penal Reform, Carceral Expansion and Greenwashing. In N. South, & A. Brisman (Eds.), *Routledge International Handbook of Green Criminology* (pp. 260–276). Routledge.

McGlynn, C., & Westmarland, N. (2019). Kaleidoscopic Justice: Sexual Violence and Victim-Survivors' Perceptions of Justice. *Social & Legal Studies, 28*(2), 179–201. https://doi.org/10.1177/0964663918761200

Mohai, P., Pellow, D., & Timmons Roberts, J. (2009). Environmental Justice. *Annual Review of Environment and Resources, 34*, 405–430. https://doi.org/10.1146/annurev-environ-082508-094348

Natali, L., & Hall, M. (2021). A Green Criminological Approach to Environmental Victimization and Reparation: A Case for Environmental Restorative Justice. In L. Centemeri, S. Topçu, & J. P. Burgess (Eds.), *Rethinking Post-Disaster Recovery* (pp. 165–184). Routledge.

Pečar, J. (1981). Ekološka Kriminaliteta in Kriminologija. *Revija Za Kriminalistiko in Kriminologijo, 34*, 33–45.

Pellow, D. N. (2004). The Politics of Illegal Dumping: An Environmental Justice Network. *Qualitative Sociology, 27*(4), 511–525. https://doi.org/10.1023/B:QUAS.0000049245.55208.4b

Quinney, R. (1970). *The Social Reality of Crime*. Little, Brown and Company.

Rodríguez Ramos, L. (1981). Presente y Futuro de la Protección Penal del Medio Ambiente en España. In CEOTMA/MOPU (Ed.), *Derecho y Medio Ambiente* (pp. 283–316). CEOTMA/MOPU.

Short, D. (2016). *Redefining Genocide: Settler Colonialism*. Zed Books.

Sollund, R. (2021). Green Criminology: Its Foundation in Critical Criminology and the Way Forward. *The Howard Journal of Crime and Justice, 60*(3), 304–322. https://doi.org/10.1111/hojo.12421

Sollund, R. (2023). Humans, Nature and the Other Animals: Developments in Green Criminology in Northern Europe. *Newsletter of the European Society of Criminology*, https://esc-eurocrim.org/v2/wp-content/uploads/2023/11/Newsletter_21-02_2023.pdf

South, N. (2014). Green Criminology: Reflections, Connections, Horizons. *International Journal for Crime, Justice and Social Democracy, 3*(2), 5–20. https://search.informit.org/doi/epdf/10.3316/informit.270196734841610

Varona, G. (2021). Why an Atmosphere of Transhumanism Undermines Green Restorative Justice Concepts and Tenets. *International Journal of Restorative Justice, 4*(1), 41–59. https://doi.org/10.5553/TIJRJ.000066

Weis, K., & Borges, S. S. (1973). Victimology and Rape: The Case of the Legitimate Victim. *Issues in Criminology, 8*(2), 71–116.

White, R. (2008). *Crimes Against Nature: Environmental Criminology and Ecological Justice*. Routledge.

White, R. (2013). *Environmental Harm: An Eco-Justice Perspective*. Policy Press.

White, R. (2016a). Eco-Justice and the Moral Fissures of Green Criminology. In J. Jacobs, & J. Jackson (Eds.), *The Routledge Handbook of Criminal Justice Ethics* (pp. 76–94). Routledge.

White, R. (2016b). Four Problems for Specialist Courts in Dealing with Nonhuman Environmental Victims. In T. Spapens., R. White, & W.

Husman (Eds.), *Environmental Crime in Transnational Context: Global Issues in Green Enforcement and Criminology* (pp. 139–153). Routledge.

White, R. (2017). Reparative Justice, Environmental Crime and Penalties for the Powerful. *Crime, Law and Social Change, 67*, 117–132. https://doi.org/10.1007/s10611-016-9635-5

White, R. (2018). Green Victimology and Non-Humans Victims. *International Review of Victimology, 24*(2), 239–255. https://doi.org/10.1177/0269758017745615

White, R., & Kramer, R. C. (2015). Critical Criminology and the Struggle against Climate Change Ecocide. *Critical Criminology, 23*(4), 383–399. https://doi.org/10.1007/s10612-015-9292-5

Wyatt, T. (2013). *Wildlife Trafficking: A Deconstruction of the Crime, the Victims, and the Offenders*. Springer.

2

Constructing an Eco-Justice Framework

Rob White

1 Introduction

This chapter provides a detailed exposition of the main elements of an eco-justice framework. This consists of specific justice notions that are integrated into a single overarching conception of justice. The eco-justice framework comprises three key dimensions: environmental justice (humans), ecological justice (ecosystems) and species justice (non-human animals and plants). Consideration is also given to a recent addition, namely, extraterrestrial justice (cosmological entities).

I have personally been developing this framework for some time, and the chapter provides an opportunity to flesh it out in greater detail. It begins by discussing the origins of the eco-justice framework, one that is specifically linked to green criminology, and its evolution from inception to the present day. The framework initially emerged out of discussions of eco-philosophy as well as analysis of diverse environmental

R. White (✉)
University of Tasmania, Hobart, TAS, Australia
e-mail: r.d.white@utas.edu.au

R. Killean et al. (eds.), *Conceptualising Justice*, Palgrave Studies in Green Criminology, https://doi.org/10.1007/978-3-032-19791-7_2

and animal rights movements. From these beginnings, the framework has over time incorporated plants, and more recently cosmological bodies, into its broad perspective.

The main part of the chapter discusses the three major components, and briefly the fourth, of the eco-justice framework with the intention of explicating its constitutive elements and ideas. It begins by reviewing the origins and ongoing development of the framework. It then turns to specific detailed consideration of environmental justice, ecological justice, species justice and extraterrestrial justice, concluding with a general summary. Given space limitations, the chapter does not discuss the tensions within and between each specific dimension of eco-justice, nor evaluate the framework's overall strengths and limitations (instead, see White, 2013). The necessity for and complexity of weighing up diverse rights and interests in concrete situations can nonetheless be acknowledged and is discussed in greater depth elsewhere (White, 2022).

2 Framework Origins and Development

The broad orientation of green criminology is directed at exposing instances of environmental harm including harms other than those specified in criminal law (Brisman & South, 2021). Importantly, an essential aspect of green criminology is that the moral community ought to be extended to include non-human nature. This involves a new politics of obligation, one that includes obligations to animals, trees, mountains, oceans and other members of the non-human biotic and abiotic community (Smith, 1998). This notion of moral community thus emphasises human obligations to all living things (and, indeed, to the non-living such as rivers and mountains), and the need to carefully assess the impacts of human activity across the human and non-human domains. It also incorporates notions of risk and precaution to minimise potential harm.

Harm (e.g. assessments of damage and destruction) and morality (e.g. conceptions of right and wrong) are conceptualised in an eco-justice perspective as incorporating concern about the rights and well-being of humans, ecosystems and species. Justice is thus construed as referring to fairness, inclusion and equity in regard to both the human and the non-human. It asks the question 'justice for whom?' and then attempts to distill the principles that can guide us in dealing with harmful practices and conflicting situations. Eco-justice wishes to establish a moral basis for action as well as the institutional measures to support this. It deals with issues holistically in ways that attempt to maximise ecological sustainability and good environmental outcomes. An eco-justice perspective asserts that the fate of humans is inextricably bound up with that of ecosystems and the varied species that inhabit these.

While the framework has taken shape over the past twenty years or so, the underpinning ideas of eco-justice, as discussed herein, are not new. For example, there is a longstanding argument that the inherent interests of 'natural objects' ought to be protected through legal actions by the objects themselves, with humans serving as their guardians or trustees (Stone, 1972). Similar arguments are put forward in relation to rights for non-human animals. At its core, justice is about rights and duties, and in particular the rights to 'be' and to 'do' (Cullinan, 2003). A central contention of eco-justice is that such rights, in some form, ought to be attributed to (or in some interpretations, conferred upon) natural objects such as rivers and mountains, and animals at both species-level and as individual creatures, as well as plants. Moreover, the lens of eco-justice can be used to shine light on various forms of victimisation, including victims of environmental crime that span the human and the non-human (White, 2022).

My original eco-justice framework consisted of just two dimensions: environmental justice (humans) and ecological justice (ecosystems) (White, 2007). In this iteration, environmental justice was construed in relation to human-centred notions and the uneven distributions of environmental benefits and harms. Ecological justice, on the other hand, was focussed on the harms of destructive human activities impacting on specific ecosystems. Discussion of these concerns, at this time, was profoundly shaped by philosophical debates involving 'deep ecology' proponents and those advocating for a 'social ecology' approach (Bookchin, 1988; Mathews, 1987; Tokar, 1988). Aside from persistent definitional difficulties (due to constantly shifting interpretation of these terms), there was a tendency for one strand of deep ecology to treat humanity as an undifferentiated whole and to view humans solely as being the biggest threat to nature. There was also the view that any modification of nature to serve human well-being was something to be condemned. By contrast, social ecology emphasised humanity and the ability of humans to change themselves and the natural world. This perspective spoke to the historical unity of ecological and social concerns and sought to identify and combat the social relations of hierarchy and domination that are directly linked to the exploitation of nature (White, 1994). This unity is what was also reflected in my initial two-pronged conceptualisation of justice.

These eco-philosophical ideas were further fleshed out, for criminological purposes, in an article which elaborated on anthropocentric, biocentric and ecocentric perspectives on humans and the natural environment and then applied these perspectives to the analysis of environmental harms pertaining to the clear felling of old growth forests (Halsey & White, 1998). This article criticised anthropocentrism, which views humans as superior to all other entities (thereby privileging self-interest, human benefit and the instrumental use of the natural world), and biocentrism, which views the non-human as having intrinsic value and humans as (simply) morally and ethically equal to all other entities (thus opening the door to misanthropic portrayals of human disaster; for example, AIDS was lauded by some as a solution to over-population). The favoured third option, ecocentrism, views humanity as part of nature but also as having a responsibility that extends to both human and non-human life. This interpretation of ecocentrism attempts to strike a balance between the instrumental and intrinsic conceptions of non-human nature put forward by anthropocentrists and biocentrists respectively. That is, 'while there is an explicit recognition that human beings—as living, breathing components of the natural world—need to impact upon or utilise non-human nature in order to survive, there also exists the realisation that human beings need—in the interests of future generations and the well-being of non-human nature—to develop ecologically sustainable ways of satisfying their basic needs' (Halsey & White, 1998: 356). These sentiments translated later into the interrelated notions of environmental (human) and ecological (ecosystem) justice within the eco-justice framework.

The third strand of the eco-justice framework, pertaining to species justice, began with the increasing academic attention given to the status of, and harms to, non-human animals. This was partly due to the efforts of green criminology-aligned writers to treat 'animal' issues seriously as part of the emergent environmental harm discussions (Beirne, 2007). However, this is only part of the picture. Consideration of species justice also emerged out of analysis of diverse environmental and animal rights movements (White, 2013). For example, it was noted that the Environmental Justice Movement was concerned with human communities affected by harmful pollution and toxic waste. In the United States, this victimisation was linked to environmental racism (Bullard, 2005). The Wilderness Society, by contrast, centred its campaigns on protecting old growth forests in Tasmania, with activities informed by ecocentric perspectives that value nature for its own sake. On the other hand, groups such as Greenpeace have long been motivated by protection of non-human species such as whales. Each type of social movement tended to focus on specific categories of victims. Together, they cover humans as well as non-human entities such as animals, rivers and trees. This inclusiveness means that my research attention was increasingly directed to matters of habitat, species and landscape, as well as human communities.

Recognition of these interconnections led to the first systematic review of eco-justice as a three-pronged framework (White, 2013). It was during this in-depth consideration of the principles and approaches of eco-justice that the question of 'what about plants?' was raised (White, 2013: 153). The argument was that issues of biodiversity and protection from destruction by humans (and indeed other non-human animals) also extend to species other than animals. For example, the species covered in the appendices of the Convention on International Trade in Endangered Species of Wild Fauna and Flora (CITES) include plants. Indeed, the concept of 'wildlife' in CITES incorporates both fauna (non-human animals) and flora (plants). Rather than rely on the concept of speciesism, however, the eco-justice framework began to increasingly consider the importance of biodiversity when it came to plant life (although biodiversity is also of major concern to non-human animals).

Over time, the eco-justice framework has incorporated plants, and most recently, cosmological entities (for example, the Moon), into its broad perspective (Lampkin & White, 2023). Each of these areas has also had a significant social movement exemplar. For instance, some organisations concentrate on so-called 'brown' issues and are concerned with the pollution of air, land and water. Other activists focus on 'green' issues that deal with conservation and preservation of wilderness areas and issues such as deforestation and illegal logging. Groups are also involved in campaigns against the use of technologies such as genetically modified organisms in plants and non-human animals. And animal welfare and animal rights groups continue to be vocal in their condemnations of animal cruelty and call for better conditions and respect for the non-human animal. Meanwhile, recent statements regarding the 'Rights of the Moon' mirror those pertaining to the rights of nature here on Earth (Australian Earth Laws Alliance, 2021). Out of this cauldron of ideas and activism has emerged a more sophisticated and complex eco-justice framework.

3 An Eco-Justice Framework

As mentioned, there are three key components to the eco-justice framework, with the recent addition of Outer Space considerations providing a tentative new dimension. This section provides a summary of the main elements of the eco-justice framework as it has evolved to date, bearing in mind that the framework as a whole is constituted by these elements.

3.1 Environmental Justice

The environmental justice strand of eco-justice essentially centres on humans and human communities. Environmental rights are seen as an extension of human or social rights, and are related to the quality of human life, now and into the future. This dimension of eco-justice largely stemmed from analysis of the work of the environmental justice movement (EJM) in the United States. The EJM emphasised that those individuals and communities who lived in the worst environmental areas and most polluted parts of cities were overwhelmingly comprised of poor and marginalised people, and frequently were people of colour (Bullard, 2005). The objective of the EJM was to counter environmental discrimination and racism, and to advocate for better environmental conditions as part and parcel of a social justice agenda.

Looked at broadly from a green criminology perspective, there are two categories of victim—universal victims, and specific victims (White, 2011). Some environmental harms, such as global warming, affect all humans regardless of class, race, ethnicity, age and other ascribed and achieved characteristics. On the other hand, poor environmental conditions, as manifest in polluted waters, land and air and geographical areas with high levels of ecological destruction (e.g. deforestation) and contamination (e.g. filthy rivers) tend to be concentrated in areas populated by the less powerful. This is a global phenomenon. Discrimination and poverty (and vice versa, dominance and wealth) feed into an unequal and uneven distribution of environmental goods, such as freshwater, fresh air, arable land and adequate sanitation. Colonialism and contemporary efforts to harvest natural resources in territories outside the metropoles of the global North (and the elite centres of the global South) mean that Indigenous and traditional communities are particularly vulnerable to exploitation, and exclusion and denial of environmental services and goods, all of which also affect efforts to maintain ecological sustainability (Crook, 2024; Goyes, 2019).

Striving for justice involves both *intra*generational and *inter*generational equity. The first refers to the notion that every person, as a human right, should have access to clean and safe environments regardless of socio-economic background, gender, (dis)ability, nationality, race and so on. The environment is a social good that ought to be universally available and shared equally amongst the human population. This abstract right needs to be accompanied by concrete efforts to address legacy issues such as colonial subjugation, and slavery. Rights of collectivities, such as Indigenous peoples, as well as individual human rights need to be recognised, respected and accompanied by dedicated legal protections. Compensation, both material in the form of land grants and financing, and opportunities in regard to jobs, welfare, health and education, are needed to address current ills and historically created wrongs. This also extends to historical patterns of inequity at a global level, particularly when considering the causes and consequences of climate change (White, 2018).

On the other hand, the idea of intergenerational equity looks to the past and future as benchmarks for environmental well-being (Weiss, 1992, 2008). As a temporal concept, it incorporates past generations, present generations and future generations. Members of the present generation hold the Earth in trust for future generations, while at the same time they are beneficiaries of its resources. Equity must flow to present generations from past generations, while simultaneously present generations must ensure that equity flows to future generations. Moreover, the dynamics of nature (both human and non-human) demand attention to the vagaries of change that naturally occur over time.

The temporal nature of intergenerational equity means there are inherent links between intergenerational equity and intragenerational equity. This is because the obligation is for each generation to pass the planet on in no worse condition than it received it, and to provide equitable access to its resources and benefits. The planet marks out the scale of the obligation. Therefore, equality of rights extends in time as well as space to embrace both generations and geographical regions (and the people who inhabit these). Intergenerational planetary rights may be regarded as group rights, as distinct from individual rights, in the sense that generations hold these rights as groups in relation to other generations—past, present and future (Weiss, 2008). That is, these are 'generational rights' which must be conceived in the temporal context of generations, rather than rights of identifiable individuals (although there are identifiable interests of individuals that the group rights protect). These generational rights can be evaluated by applying objective criteria and indices to the planet from one generation to the next.

The environmental justice dimension of eco-justice therefore tends to put the emphasis on specific oppressed or exploited groups and frequently refers to allied concepts such as environmental racism and environmental inequality (Agyeman & Carmin, 2011; Bullard, 2005). The quest for environmental equity alludes to attempts to reduce hazards, risks and harms for all sections of the population, now and into the future, and to ensure proportionate enjoyment of 'clean and green' environmental amenities. It involves avoiding discriminatory access to clean air, water and soil as well as holding to account companies and governments that damage environments in ways that negatively affect specific population groups and individuals. Environmental harm is constructed in relation to human-centred notions of value and use.

3.2 Ecological Justice

The notion of ecological justice encapsulates concern with both the biotic and the abiotic. An ecosystem is made up of both abiotic components (air, water, soil, atoms and molecules) and biotic components (plants, animals, bacteria and fungi) (Merchant, 2005). Harms that affect non-living entities such as rivers, mountains and oceans and living entities that include humans as well as non-human animals, plants and ecosystems, are the focus of ecological justice. Consideration of the non-human environmental entity incorporates discussion of individual landscape features, specific living entities and particular ecosystems.

This approach views humans as one component of complex ecosystems that should be preserved for their own sake. When destruction, degradation or diminishment of these occurs, then sustained analysis of the nature of the harm needs to take place. How the nature-human interface occurs is important and can be evaluated from the point of view of ecological sustainability. Ultimately, humans need to engage in stewardship and custodianship of rivers, mountains and forests—diverse ecosystems and ecological communities—and thereby exercise a form of ecological citizenship centred on nature.

The centrepiece for action is the non-human, rather than humans and human interests. For instance, Cullinan (2003: 118) points out that 'a fundamental river right (i.e. the riverine equivalent of a human right) would be the right to flow. If a water body couldn't flow it wouldn't be a river, and so the capacity to flow (given sufficient water) is essential to the existence of a river'. Therefore, 'from the perspective of the river, building so many dams across it and extracting so much water from it that it ceased to flow into the sea, would be an abuse of its Earth rights' (Cullinan, 2003: 118). The overarching concern is with the quality of the planetary environment (that is seen to possess its own intrinsic value) and the rights of animal and plant species to flourish in this habitat. Analysis in this instance is frequently directed at environmental harm linked to specific ecosystems and endangered ecological communities.

In this perspective, the notion of sustainable development is interpreted as reflecting the anthropocentric idea that the instrumental use of nature needs to be carried out in ways that ensure future exploitation as well as present entitlements (Bosselmann, 2010; De Lucia, 2014). This can be contrasted with ecological sustainability that refers to ecosystem needs rather than human needs. In the former, the strategic choices revolve around how best to manage natural resources in ways that help to alleviate outcomes such as loss of biodiversity, pollution of air and water and land degradation because otherwise the whole system will be put into jeopardy. In practice, the tendency is to ignore these considerations altogether. In the latter, the emphasis is on intrinsic values and protection of non-human nature for its own sake. This is sometimes expressed in the language of 'preservation' which implies a severing of human contact and un-interfered-with natural status quo, but more usually it alludes to conservation values which encapsulate the notion of continuous nature-human interaction that nonetheless maintains the basic ecological integrity of a region.

What necessarily follows from an ecological perspective are notions of interconnectedness and human obligation to the non-human world around them. Justice demands that how humans interact with their environment be evaluated in relation to potential harms and risks to specific creatures and specific locales as well as the biosphere generally. A central concept is ecocentrism. This refers to the view that the environment ought to be valued for its own sake apart from any instrumental or utilitarian value to humans. All lifeforms have an intrinsic value, and even the non-living (i.e. rocks, river waters) contribute to the greater whole. A fundamental aspect of ecocentrism is that it views non-human entities, such as animals, plants and rivers, as potential rights-holders and/or objects warranting a duty of care on the part of humans because non-human environmental entities' interests are seen as philosophically significant—that is, deserving greater respect and formal recognition by humans than hitherto has been the case (Schlosberg, 2007). Ecocentrism asserts that humans are just one part of a wider community of subjects and that the well-being of each member of the community is reliant on the well-being of Earth as a whole.

A perspective based on ecology mirrors but is not the same as that based on Indigenous cosmology (Graham, 1999; Goyes et al., 2021). While each values nature and sees the web of life as forged in interconnectedness, reciprocity and sustainability, Indigenous worldviews incorporate the notion that all entities—biotic and abiotic—are life forces. The river, the mountain, and the Earth itself are all alive. In many cases, they are also inseparable from their local human counterparts, as expressed for example in Māori connections to the flowing freshwater of rivers. This is encapsulated in the common tribal saying, 'I am the river and the river is me' (Morris & Ruru, 2010). In other words, identity is merged in the unity of the human and the non-human.

Different laws set out different kinds of human obligations toward the natural environment. A 'rights of nature' approach places emphasis on the status and legal standing of the non-human (Maloney & Burdon, 2014; O'Donnell & Talbot-Jones, 2018). An 'ecocide' approach, however, is primarily concerned with preventing harms to the environment (Higgins, 2012). This describes an attempt to criminalise human activities that destroy and diminish the well-being and health of ecosystems and the species within these, including humans. Ultimately, these various legal initiatives converge in attempting to provide a legal basis for enhanced protection of the environment in its own right (Killean & Newton, 2025).

Acknowledgement of the intrinsic worth and value of nature is vital from the point of view of an ecocentric approach, but this does not mean that instrumental uses of nature by humans are thereby rendered unimportant (Donnelly & Bishop, 2007). The instrumental use of nature, however, can nonetheless be guided by ecocentric considerations, such as doing the least amount of harm in the process of doing so. Humans, too, are part of nature, and this is reflected in how we interact with and use natural resources. In accordance with ecological justice, humans have a duty of care to protect, conserve and preserve the natural environment. This is indicated in support for notions such as the 'rights of nature' and as manifest in policies such as establishing national parks and banning logging in old growth forests. The role of humans is to provide stewardship and custodianship of rivers, mountains and forests and thereby exercise a form of ecological citizenship centred on nature. The ecological justice contribution to the eco-justice framework is to emphasise ecology and ecosystems.

3.3 Species Justice

Species justice is based on the idea that non-human animals have an intrinsic right not to suffer abuse and that speciesism is inherently discriminatory. The matter of animal rights and animal welfare is encapsulated in that work concerned with species justice (Beirne, 2007). Speciesism refers to the practice of discriminating against non-human animals because they are perceived as inferior to the human species in much the same way that sexism and racism involve prejudice and discrimination against women and people of different colour (Ryder, 2010). Animal rights supporters argue that anthropocentric discourses construct two distinctive kinds of animals—human and non-human—and that to some extent both have rights and interests as sentient beings; they believe, however, that the dominant ideology of speciesism enables humans to exploit non-human animals as commodities to be eaten, displayed, hunted and dissected for human benefit.

Animals are categorised and utilised by humans in different ways that range from factory farming through to laboratory animals. Depending upon human use, animal welfare and rights are protected differentially depending upon species and circumstance (White, 2011, 2013). Animal protection laws operate strongly when it comes to companion animals, for instance, but are much less restrictive in the case of pigs grown for eventual slaughter. It has been observed that non-human animals are frequently considered in primarily instrumental terms (as 'pets', as food, as resources) in mainstream and conservation criminology, and are categorised in mainly anthropomorphic terms (such as 'wildlife', 'fisheries') that belie the ways in which humans create and classify animals as Other (Beirne, 2009).

Within the broad animal concern paradigm, the focus of the welfarist approach is the humane treatment of animals (Fraser, 2010). This model advocates for the protection of animals through increased welfare-based interventions but not the prohibition of animal exploitation. Implicit in this is that animals may still be exploited for their flesh, fur and skin provided their suffering is not 'unnecessary', or as often put, the animals are treated humanely. At the other end of the spectrum is the rights-based approach. This approach argues for the abolition of animal exploitation through both legal and non-legal change and for the legal recognition of rights for animals. Central to this approach is changing the legal character of animals from property to legal, rights-bearing entities (Francione, 2008; Regan, 2010). How harm to non-human animals is conceptualised thus very much depends upon the perspective one has on the ontological status of animals (their essential 'nature' or 'being'), and how one views the relationship between humans and non-human animals.

Another aspect of species justice is the desire to protect species from extinction. This is related to but different from concerns about abuse and exploitation as such. For example, the International Union for the Conservation of Nature has a Red List that categorises the survival status of species, ranging, for example, from 'extinct', 'critically endangered' and 'near threatened' to 'least concern' (IUCN, 2025). Considerations of the welfare and rights of non-human animals therefore include existential questions pertaining to the very survival of species on the planet.

Recent work has also argued for the further extension of non-speciesist green criminology to include invertebrates (Durrant, 2024). These include insects, crabs, octopuses, snails, earthworms, sea anemones and many more species. It is observed that these invertebrate species make up over 95% of animal life. They are essential to ecosystems and form a vital part of the web of life. Accordingly, they, too, should be considered within an eco-justice framework.

For plants, the idea is that they should not experience the degradation of habitat to the extent that threatens biodiversity loss. Rather than drawing upon the language of speciesism, therefore, protection of plants is generally framed as a matter of biodiversity. This concept refers to the different plants, animals and micro-organisms, and their genes that together make up life on the planet. It also includes reference to the terrestrial (land), marine (ocean) and freshwater (inland water systems) ecosystems of which they are a part. With biodiversity, the key ecological message is 'the more the merrier'—the greater the number of species, the greater the resilience of the system to potential catastrophe, whether this is fire, drought or climate change (Secretariat of the Convention on Biological Diversity, 1992, 2010).

The protection, maintenance and extension of biodiversity are important in ecological terms. Such concerns are also reflected, for example, in law enforcement efforts relating to trade in endangered species (e.g. the Convention on International Trade in Endangered Species of Wild Fauna and Flora) as well as a range of other international agreements protecting specific environments (e.g. Convention on the Conservation of Antarctic Marine Living Resources) and specific species (e.g. International Convention for the Regulation of Whaling) (White, 2011).

Many different forces and factors are involved in potential biodiversity loss with respect to plants. For example, bioprospecting (the worldwide search for plants with special properties, such as for medicinal use) that is unregulated and inappropriate can be considered a form of overexploitation which has the potential to degrade ecosystems and increase biodiversity loss, as well as impacting on the rights of the communities and states from which the resources are taken (Mgbeoji, 2006). Probably the biggest threats to biodiversity are those associated with contemporary agricultural methods related to the corporatisation of agriculture (Gray & Hinch, 2018). This has been accompanied by significant changes in land use, such as a major expansion in cropland worldwide, and greater reliance on fewer crop species as well as application of genetic modification technologies. This has resulted in a tendency toward monoculture, since uniformity means ease of cultivation and harvest, and higher yield, which translates into higher profit. Moreover, intensive use of land and soils that rely upon chemical additives to ensure productivity, rather than, for example, traditional methods of crop rotation, further diminishes longer term agricultural viability.

The species justice perspective incorporates concerns about animal rights and welfare, acknowledgement of the status and role of insects, crustaceans, molluscs and other invertebrate species, and the importance of protecting vulnerable and threatened vegetation. The emphasis is on protecting fauna and flora. In regard to non-human animals, a concern is to counter 'speciesism', that is, the discriminatory treatment of non-human animals and invertebrates. For plants, the focus is on biodiversity and reducing threats to this. Environmental harm is constructed in relation to one-on-one harm (e.g. killing individual animals), institutionalised harm (e.g. factory farms) and harm arising from human actions that affect climates and environments on a global scale (with negative consequences for plant life and animals). The threat of extinction looms large for both animal and plant. Such concerns form the basis of the species justice contribution to the eco-justice framework.

3.4 Extraterrestrial Justice

'Outer Space' refers to the expanse that exists beyond Earth and its atmosphere and between celestial bodies. Other common terms in this area include 'extraterrestrial'. The notion of extraterrestrial justice refers to the idea that Outer Space or extraterrestrial bodies such as moons, asteroids, planets and comets have their own special character and social and cultural status, and that this should be preserved as far as possible in any interactions involving humans (Lampkin & White, 2023). Caution should be taken in how humans interact with and exploit these bodies insofar as economic value should not trump other kinds of values such as intrinsic value and the right to 'be'.

This discussion of justice is less about identifying a specific new dimension than expressing a new *application* for the eco-justice framework. Nonetheless, it is presented here as an additional dimension to the eco-justice framework for two reasons: first, to highlight the importance of off-planet entities as worthy of moral attention; and second, because there are unknowns in this domain that warrant hesitation and careful assessment—there is a universe out there which requires further broadening of our knowledge and understanding about space, time, matter and what lies in-between.

Harm in the context of Outer Space may parallel crimes on the Earth (e.g. theft, environmental destruction) but is also space-specific insofar as it may stem from and relate to endeavours that only take place in

Outer Space (e.g. space junk). Space junk may include the proliferation of materials in near-Earth space such as satellites, some of which fall to Earth. It also includes Earth-made products left on other celestial bodies, such as land rovers left on the Moon and Mars, although these may be deemed valuable heritage items rather than junk per se. Earth-based production for space-related activities involves generation of pollutants and contamination of land, air and water. In space, asteroids and moons may be exploited for their minerals in ways that fundamentally transform them. Methods and techniques of natural resource extraction and use in Outer Space may involve exploitation of humans and nature, with unforeseen consequences for both.

Similar in many respects to the rights of nature movement on Earth, there is emergent advocacy for extraterrestrial rights of nature. So far, this has mainly taken the form of a manifesto for the 'Rights of the Moon' (Australian Earth Laws Alliance, 2021). The manifesto expresses ideas and interests that encompass the spiritual and the material, the known and the unknown, time and space, ecology and cosmology. It concludes by stating that:

The Moon possesses fundamental rights, which arise from its existence in the universe, including:

- the right to exist, persist and continue its vital cycles unaltered, unharmed and unpolluted by human beings;
- the right to maintain ecological integrity;
- the right to be defined as a self-sustaining, intelligent, cohesive, intact lunar ecosystem, beyond current human comprehension;
- the right to independently maintain its own life-sustaining relationship with the Earth's environments and living creatures; and
- the right to remain a forever peaceful celestial entity, unmarred by human conflict or warfare (Australian Earth Laws Alliance, 2021).

To date, green criminology has extended the typically human-centred conception of justice to include ecosystems and plant and animal species. Conventional crimes involving humans as victims are also still included in the broad eco-justice framework, but the framework places strong emphasis on the non-human. The extraterrestrial justice perspective suggests that analysis of human interventions in Outer Space must also account for non-human interests and entities, although in this case it is celestial bodies that are most relevant, and the possibility of non-Earth lifeforms is also a consideration.

Of particular concern to the extraterrestrial justice perspective are issues pertaining to the commercialisation and privatisation of space endeavours, the militarisation of space activity including dual purpose technologies (scientific and military), and the lack of adequate regulation of space junk (including adding even more to the mess circulating above the planet). These trends portend serious harms occurring in Outer Space that may well affect humans on Earth as well. Justice and harm considerations also apply to specific cosmological entities such as asteroids, moons and comets, and include apprehension about the spread of debris and contamination across vast distances.

The question of extraterrestrial *life* of course also raises additional complications with regard to eco-justice. Much depends upon the sophistication of the lifeform, the ways in which it relates to or threatens humanity, whether it is intelligent or conscious, plant-related or animal-related and able to communicate with humans. All this depends, of course, on whether there are other life forms in the vicinity of Earth and its local planetary system, or indeed in the universe: 'Either proposition—that there is 'other life' or that there is no 'other life'—is astonishing and deeply unsettling. If the former, then there will be much yet for space criminology to consider philosophically and practically' (Lampkin & White, 2023: 15).

What scope there is for democratic input into decisions about the extraterrestrial, including the Moon, is of the highest concern as well. Space flight and space technology are increasingly in the hands of private entrepreneurs. What happens in Outer Space likewise appears to be replicating what the billionaires on Earth are doing in regard to the exercise of their power in pursuit of specific sectional interests (White, 2024). Movements such as the Moon rights movement emphasise caution, conservation and consultation. From the point of view of eco-justice, these are vital elements in establishing a moral universe in which new forms of environmental harm—including that pertaining to Outer Space—are adequately recognised and addressed.

4 Summary and Conclusion

The major foci and propositions of the four components of the eco-justice framework are presented in Table 1. Tensions exist within and across these justice approaches. This is illustrated by disputes between animal welfare and animal rights activists. It is also apparent in discussions of ecological justice, in which there are disagreements over whether humans are part of nature or apart from nature and, concretely, whether it is sensible or not to talk about preservation or conservation. These tensions, contradictions and paradoxes are discussed in greater detail elsewhere (White, 2013), but they continue to warrant more systematic scrutiny.

Further attention is needed as well in forging answers to questions pertaining to:

- *conflicts over rights* (e.g. intrinsic versus instrumental; hierarchies of rights in relation to humans, non-human species and ecosystems);
- *measuring harm* (e.g. thresholds for determining extent of harm and when to intervene and why; scale at which harm occurs, from individual to landscape and species levels); and
- *voice* (e.g. experts and advocates; democratisation and inclusion of diverse interests and viewpoints).

An ecological conception of humans within the web of life demands new ways of understanding the world around us. In support of this, the eco-justice framework puts harm and morality at its centre and asserts that our future as humans is inextricably tied to that of the non-human. It is a framework for analysis and interpretation. It is also a call to action.

Table 1 An eco-justice framework

Environmental Justice	Ecological Justice	Species Justice	Extraterrestrial Justice
Theories and Approaches			
Focus on human communities Intragenerational inequality involving discrimination against certain communities Intergenerational equity that takes into account past, present and future states of the environment	Focus on habitats and ecosystems Ecocentrism incorporates considerations of rights of nature and duty of care Ecocide refers to widescale destruction and diminishment of the environment	Focus on non-human animals, invertebrate animals, and plants Animal welfare and animal rights approaches Biodiversity and prevention of extinction	Focus on the extraterrestrial and objects or beings beyond the boundaries of the planet Concern is with cosmological bodies such as moons, asteroids, planets and comets Justice incorporates notions such as rights of the Moon
Case Studies			
Love Canal, USA Bhopal, India Abidjan, Ivory Coast	Toxic chemicals Climate change Plastic in oceans	Whales Live export of sheep Genetic modification of plants	Space junk Debris from rocket launches Mining of asteroids

(continued)

Table 1 (continued)

Environmental Justice	Ecological Justice	Species Justice	Extraterrestrial Justice
Measuring Harm and Value			
Environmental racism and evidence of discrimination Universal victimisation in which everyone is affected Select victimisation associated with vulnerability such as children and people with disabilities, and geographical regions	Estimates of direct damage, immediate and potential impact, complexity and totality of ecological damage Activities associated with destruction of selected landscapes and marine areas Re-establishment time and whether the damage can be redressed and how	Speciesism as discrimination against non-human animals Status of species as vulnerable and threatened Damage, destruction and modification of specific plants, animals and habitats	Regard for ontological status of cosmological bodies such as the Moon Contamination of land, air and water due to space-related activity on Earth Extent of waste and junk on celestial bodies and in Outer Space
Specific Social Movements			
Toxic Action Network Environmental Justice movement Clean air activists	Earth First Sierra Club Wilderness Society	Greenpeace Sea Shepherd Animal Liberation Movement	Space Criminology Group Australian Earth Laws Alliance Outer Space Institute
Critiques and Contradictions			
Not accounting for social differences in victimisation, such as gender differences Environment issues seen as part of a global pattern or particularistic emphasis on local communities, defined geographically and/or socially	Prioritises places over people Wilderness and pristine areas are in fact inseparable from nature-human interactions over time	The interests and rights of animals prioritised over humans and environments Ignoring the historical evolution of animal-human interactions	Space exploration heavily reliant on private investment and entrepreneurs versus public spending and state-based regulation and control Militarisation of space and use of anti-satellite weapons to enforce space dominance Democratisation of space, and use of space for peaceful purposes

References

Agyeman, J., & Carmin, J. (2011). Introduction: Environmental Injustice Beyond Borders. In J. Carmin & J. Agyeman (Eds.), *Environmental Inequalities Beyond Borders: Local Perspectives on Global Injustices* (pp. 1–16). The MIT Press.

Australian Earth Laws Alliance. (2021). *Declaration of the Rights of the Moon.* AELA. www.earthlaws.org.au

Beirne, P. (2007). Animal Rights, Animal Abuse and Green Criminology. In P. Beirne & N. South (Eds.), *Issues in Green Criminology: Confronting Harms Against Environments, Humanity and Other Animals* (pp. 55–85). Willan Publishing.

Beirne, P. (2009). *Confronting Animal Abuse: Law, Criminology, and Human-Animal Relationships.* Rowman & Littlefield Publishers.

Bookchin, M. (1988). Social Ecology Versus Deep Ecology. *Socialist Review, 88*(3), 11–29.

Bosselmann, K. (2010). Losing the Forest for the Trees: Environmental Reductionism in the Law. *Sustainability, 2*(8), 2424–2448.

Brisman, A., & South, N. (Eds.), (2021). *The Routledge Handbook of Green Criminology*. Routledge.

Bullard, R. (2005). Environmental Justice in the Twenty-first Century. In R. Bullard (Ed.), *The Quest for Environmental Justice: Human Rights and the Politics of Pollution.* Sierra Club Books, San Francisco.

Crook, M. (2024). *Capitalism, Colonisation and the Ecocide-Genocide Nexus.* University of London Press.

Cullinan, C. (2003). *Wild Law: A Manifesto for Earth Justice*. Green Books in association with The Gaia Foundation.

De Lucia, V. (2014). Competing Narratives and Complex Genealogies: The Ecosystem Approach to International Environmental Law. *Journal of Environmental Law, 27*(1), 91–117. https://doi.org/10.1093/jel/equ031

Donnelly, B., & Bishop, P. (2007). Natural Law and Ecocentrism. *Journal of Environmental Law, 19*(1), 89–101. https://doi.org/10.1093/jel/eql039

Durrant, R. (2024). *Invertebrate Justice: Extending the Boundaries of Non-Speciesist Green Criminology*. Palgrave Macmillan.

Francione, G. (2008). *Animals as Persons: Essays on the Abolition of Animal Exploitation.* Columbia University Press.

Fraser, D. (2010). Animal Welfare. In M. Bekoff (Ed.), *Encyclopedia of Animal Rights and Animal Welfare, Volume 1.* Greenwood Press.

Goyes, D. (2019). *A Southern Green Criminology: Science Against Ecological Discrimination.* Emerald.

Goyes, D., Abaibira, M., Baicue, P., Cuchimba, A., Nenetofe, D., Sollund, R., South, N., & Wyatt, T. (2021). Southern Green Cultural Criminology and Environmental Crime Prevention: Representations of Nature Within Four Colombian Indigenous Communities. *Critical Criminology, 29,* 469–485. https://doi.org/10.1007/s10612-021-09582-0

Graham, M. (1999). Some Thoughts About the Philosophical Underpinnings of Aboriginal Worldviews. *World Views Environmental Culture Religion, 3,* 105.

Gray, A., & Hinch, R. (Eds.). (2018). *A Handbook of Food Crime.* Policy Press.

Halsey, M., & White, R. (1998). Crime, Ecophilosophy and Environmental Harm. *Theoretical Criminology, 2*(3), 345–371. https://doi.org/10.1177/1362480698002003003

Higgins, P. (2012). *Earth Is Our Business: Changing the Rules of the Game.* Shepheard-Walwyn Publishers Ltd.

International Union for the Conservation of Nature [IUCN]. (2025). 'IUCN Red List of Threatened Species'. *IUCN*. Retrieved August 22, 2025, from https://www.iucnredlist.org

Killean, R., & Newton, E. (2025). From Ecocide to Ecocentrism: Conceptualising Environmental Victimhood at the International Criminal Court. *International Review of Victimology, 31*(2), 238–264. https://doi.org/10.1177/02697580241269426

Lampkin, J., & White, R. (2023). *Space Criminology: Analysing Human Relationships with Outer Space*. Palgrave Macmillan.

Maloney, M., & Burdon, P. (Eds.). (2014). *Wild Law.* Routledge.

Mathews, F. (1987). Conservation and the Politics of Deep Ecology. *Social Alternatives, 6* (4), 37–41.

Merchant, C. (2005). *Radical Ecology: The Search for a Livable World*. Routledge.

Mgbeoji, I. (2006). *Global Biopiracy: Patents, Plants, and Indigenous Knowledge.* UBC Press.

Morris, J., & Ruru, J. (2010). Giving Voice to Rivers: Legal Personality as a Vehicle for Recognising Indigenous Peoples' Relationships to Water? *Australian Indigenous Law Review, 14*, 49–57. https://www.jstor.org/stable/26423181

O'Donnell, E., & Talbot-Jones, J. (2018). Creating Legal Rights for Rivers: Lessons from Australia, New Zealand, and India. *Ecology and Society, 23*(1), 7–13. https://doi.org/10.5751/ES-09854-230107

Regan, T. (2010). Animal Rights. In M. Bekoff (Ed.), *Encyclopedia of Animal Rights and Animal Welfare* (Vol. 1). Greenwood Press.

Ryder, R. (2010). Speciesism. In M. Bekoff (Ed.), *Encyclopedia of Animal Rights and Animal Welfare* (Vol. 2). Greenwood Press.

Schlosberg, D. (2007). *Defining Environmental Justice: Theories, Movements, and Nature*. Oxford University Press.

Secretariat of the Convention on Biological Diversity. (1992). *Convention on Biological Diversity (with Annexes)*, concluded at Rio de Janeiro on 5 June 1992, registered ex officio on 29 December 1993.

Secretariat of the Convention on Biological Diversity. (2010). *Global Biodiversity Outlook 3*. SCBD.

Smith, M. (1998). *Ecologism: Towards Ecological Citizenship*. University of Minnesota Press.

Stone, C. (1972). Should Trees Have Standing?: Toward Legal Rights for Natural Objects. *Southern California Law Review, 45*, 450–487.

Tokar, B. (1988). Exploring the New Ecologies: Social Ecology, Deep Ecology and the Future of Green Political Thought. *Alternatives, 15*(4), 30–43.

Weiss, E. (1992). Intergenerational Equity: A Legal Framework for Global Environmental Change. In E. B. Weiss (Ed.), *Environmental Change and International Law: New Challenges and Dimensions* (pp. 385–412). United Nations University Press.

Weiss, E. (2008). Climate Change, Intergenerational Equity, and International Law. *Vermont Journal of Environmental Law, 9*, 615–627.

White, R. (1994). Green Politics and the Question of Population. *Journal of Australian Studies, 40*, 27–43. https://doi.org/10.1080/14443059409387164

White, R. (2007). Green Criminology and the Pursuit of Social and Ecological Justice. In P. Beirne & N. South (Eds.), *Issues in Green Criminology: Confronting Harms Against Environments, Humanity and Other Animals*. Willan Publishing.

White, R. (2011). *Transnational Environmental Crime: Toward an Eco-Global Criminology*. Routledge.

White, R. (2013). *Environmental Harm: An Eco-Justice Perspective*. Policy Press.

White, R. (2018). *Climate Change Criminology*. Bristol University Press.

White, R. (2022). *Theorising Green Criminology*. Routledge.

White, R. (2024). What on Earth is Happening in Outer Space? Questioning the Space Oligarchy. In Y. Eski & J. Lampkin (Eds.), *International Handbook of Space Crime, Criminal Justice and Ethics.* Routledge.

3

Water Justice and 'Aquatic Green Criminology'

Alexander Baird and Reece Walters

1 Introduction

Water is the lifeblood of human and non-human existence. It is indispensable to human and ecological health and fundamental to human industries, ranging from agriculture, mining, and tourism. While approximately 71 per cent of the Earth's surface is covered in water, only 2.5 per cent of the water on the planet is fresh water. Most of this is stored in polar-icecaps, glaciers, and aquifers with the rest found in surface water systems, like rivers and wetlands (World Atlas, 2018). Less than one per cent of the Earth's water is accessible and suitable for human consumption (World Economic Forum, 2025).

Fresh water is a finite resource and international governments have increasingly recognised that safe water access and sanitation should

A. Baird (✉) · R. Walters
Deakin University, Melbourne, VIC, Australia
e-mail: a.baird@deakin.edu.au

R. Walters
e-mail: reece.walters@deakin.edu.au

R. Killean et al. (eds.), *Conceptualising Justice*, Palgrave Studies in Green Criminology, https://doi.org/10.1007/978-3-032-19791-7_3

be embedded within a human rights framework (UNESCO, 2024). In 2002, for example, the UN Committee on Economic, Social, and Cultural Rights issued its General Comment No. 15, which defines the right to water as the entitlement of every person to 'sufficient, safe, acceptable, physically accessible and affordable water for personal and domestic use' (cited in UN, 2010: 1). Access to safe, clean drinking water, and sanitation was not officially recognised as a human right by the UN General Assembly until 2010 (O'Donnell et al., 2024).

However, as climate change intensifies, shifts in weather patterns and global water cycles, such as rising sea levels and extended droughts, will reduce freshwater availability. This coupled with an increasing human population and its rising demand for fresh water will put even greater pressure on already scarce water resources, exacerbating global water scarcity (UNESCO, 2024). In 2022, more than 2.2 billion people globally were without access to potable drinking water and an additional 3.2 billion people lacked access to adequate sanitation, particularly women and children living in urban slums and water-scarce areas (Michel, 2023). The accelerating growth of urbanisation is outpacing freshwater availability, exacerbating water pollution and health-related issues (UNESCO, 2024). By 2030, the UN's Sustainable Development Goal 6 seeks to 'achieve universal and equitable access to safe and affordable drinking water for all', and improvement in water quality, management and efficiency (UN Global Goals, 2025). According to the UN, however, water availability is expected to decline with approximately 40 per cent of the global population facing water scarcity by 2040 (cited in Global Water Forum, 2025).

In Australia, the most semi-arid and dry continent on Earth, freshwater supplies are incredibly scarce. This is compounded by several factors, including the over-extraction and mismanagement of fresh water, ecological disorganisation, drought, and the lack of water rights for First Nations Australians, which have impacted water security since European colonisation (Baird et al., 2024; Johnson et al., 2016; Pittock et al., 2023). First Nations' Australians are among the most impacted by water inequity—excluded and marginalised under settler-state systems of water administration—while fresh water is over-allocated and prioritised for corporate agribusiness and irrigation (Hartwig et al., 2022). This has sparked a call for water justice in Australia, to address the disparities in freshwater governance, management and distribution and 'promote both "voice" and truth-telling in relation to water' (Water Justice Hub, 2024). Professor Anne Poelina, Indigenous scholar and defender of the Martuwarra Fitzroy River, stated that 'The call for water justice must be situated within the wider frame of climate justice, which focuses on local impacts and experiences, and inequitable vulnerabilities. It can't be "business as usual", water reform is restorative justice, a matter for the greater good of all' (cited in Water Justice Hub, 2025).

The aim of this chapter is to conceptualise an understanding of *water justice* through the lens of green criminology. It draws from the eco-philosophical frameworks of environmental and ecological justice to explore water justice and its associated issues in the Murray-Darling Basin (MDB), the largest and most-connected inland river system in Australia. Management of the MDB spans five states and territories, including Queensland, New South Wales (NSW), Victoria, South Australia, and the Australian Capital Territory, and is home to more than 2.3 million people and various native flora and fauna. Commonly referred to as Australia's 'food bowl', the MDB accounts for 70 per cent of Australia's agricultural water use and provides unlimited water for stock and domestic purposes, and irrigated agriculture (Leblanc et al., 2012). Finally, it explores emerging discourses in marine justice, and proposes an aquatic green criminology devoted to diminishing harm and advancing justice across waterscapes.

2 Water Justice and Green Criminology

The term 'water justice' emerged in the academic literature in the late twentieth century, as concerns about environmental justice and human rights increasingly intersected around freshwater issues (McClanahan, 2014). It gained prominence through its connection with broader social justice movements, including poverty reduction, climate justice and Indigenous rights. These movements advocated for equal access to water, challenging oppressive state policies and the commercialisation of water by powerful corporate interests.

Water justice refers to the fair and equitable distribution, access, and management of water resources, with a focus on addressing social, environmental, and economic inequalities related to water (Hartwig et al., 2022). It is based on the principle that water is a fundamental human right and should be accessible to everyone, regardless of social status, income or geographic location, and requires multidisciplinary discourses and solutions (United Nations, 2010).

In the literature on water law and policy, water justice is often portrayed as a universal, transcendent ideal which asserts that fresh water 'is' and 'should be' available to all users, provided they align with the established hydro-social order (Boelens et al., 2018; Zwarteveen & Boelens, 2017). This means that local water users, communities, and cultures, often at the cost of their own rights and traditions, are encouraged to trust and adhere to the formal systems of the state and market that regulate and distribute fresh water, all in the name of achieving water justice. In practice, however, water justice tends to be a neoliberal, market-driven concept that serves the elitist, capitalist interests of the dominant water society, sidelining the rights, and voices of marginalised groups (Boelens et al., 2018). Marginalised groups are excluded from decision-making processes in formal water governance and distribution and consequently bear the brunt of water inequities. These inequities include cultural discrimination, ecological destruction, water contamination and health issues, all in the name of pursuing water justice (McClanahan, 2014).

Boelens et al. (2018) argue that we must deviate from this universalist, descriptive understanding of water justice towards a more grounded, relational approach that centres the lived experiences of those directly affected by water inequities. This approach acknowledges the deep, local knowledge held by communities who have long interacted with water systems, which goes unrecognised in formal decision-making processes (Hartwig et al., 2022). By grounding water justice in locally situated contexts, it can be understood not just as an abstract legal or economic principle, but as a dynamic sociocultural and ecological issue shaped by history, culture, and power (Sultana, 2018). This relational perspective requires a critical examination of water injustices, as these disparities reveal the deep-seated inequalities in how water resources are distributed, accessed, and managed.

To date, much of the existing water justice literature has been written from an environmental justice perspective, which focuses on protecting access to fresh water for drinking, recreation, sanitation, and licensed rights for commercial or industrial uses (Kallhoff, 2014). That is because fresh water is viewed primarily as an economic resource, which is extracted, commodified and sold at high prices by market forces, with human interests at the forefront of its governance, management, and protection (Johnson et al., 2016). Consequently, marginalised groups shoulder the burden of water injustice, facing unequal access to clean drinking water and the impacts of environmental racism that concentrate hazardous waste and industrial facilities in their communities (Baird, 2024).

Green criminology, which focuses on environmental harms caused by human activities, explores how water injustices result from water pollution, over-extraction, and privatisation, particularly when state-corporate interests prioritise profit over peoples' basic right to clean, accessible water and the environment (Johnson et al., 2016). We argue such water injustices, deeply rooted in socioeconomic and political structures, reflect how powerful actors, especially neoliberal states and corporations, control access to clean water, leaving others to face contamination, depletion, or scarcity (McClanahan, 2014).

While most of the water justice literature is concerned with safeguarding freshwater resources for human benefit, analysing water (in)justices through the lens of green criminology sheds light on the structural and systemic forces that perpetuate environmental degradation and social inequality in freshwater access (Baird, 2024). Water justice in green criminology also acknowledges the harms inflicted on river systems themselves, viewing them as free-flowing autonomous entities, and supports granting legal rights to rivers as part of the broader pursuit for ecological water justice (O'Donnell et al., 2024).

3 Over-Extraction, Production, and Expansion

In this section, we explore how the expansion of agricultural production, particularly cotton and almond farming, has contributed to an unequal distribution of water resources within the water market. Water markets emerged in Australia during the late twentieth century, with local farmers privately trading water entitlements between irrigation districts and neighbouring communities. Dams, channels, and weirs were constructed along inland waterways to secure a more reliable water supply to expand market operations. During this 'expansionary phase' of water market development, concerns about over-extraction began to emerge, as the growing demand for irrigation and increasing competition among users started to stretch the MDB's capacity to support both agricultural production and the ecological health of its rivers (Leblanc et al., 2012; Wheeler et al., 2014).

In 2004, the Australian government separated land and water rights across the MDB, enabling farmers to trade water licences independently of their landholdings (Australian Competition and Consumer Commission [ACCC], 2021). This shift enabled market participants to adjust their water supply, generate income by selling water rights and expand agricultural production, all while improving water efficiency and productivity. In the water market, freshwater allocations are directed to the highest-value crops to maximise its economic efficiency and ensure optimal returns for the landholder, corporation and its shareholders (ACCC, 2021). However, this often comes at the expense of ecosystems, rivers and communities that depend on these resources for survival.

In the Northern Basin, particularly northern NSW and southern Queensland, the production of broadacre crops like cotton has severely impacted the river system and its downstream communities (Grafton, 2020). Cotton was planted in Australia during the early nineteenth century with commercial production growing in Queensland and NSW in the 1960s (De Garis, 2013). Over time, innovations in irrigation technology and farming practices enhanced the efficiency of cotton production, establishing Australia as a leading global producer and exporter of cotton. Despite challenges such as drought and inflation, the Australian cotton industry has enhanced its water efficiency and productivity by 52 per cent since 1997 (Cotton Australia, 2025).

In recent decades, however, the expansion of agricultural land and growth of private water storages for cotton and other broadacre crops has contributed to an unsustainable level of water extraction from the river system (Grafton, 2020). The capacity of on-farm storages has grown by 150 per cent, increasing from 557 gigalitres in 1993 to 1393 gigalitres in 2020 (Brown et al., 2022). Most of this water is collected by intercepting and diverting floodwaters from the monsoon season into man-made storages, before it naturally recedes and flows into regulated river channels downstream. This method of water extraction (floodplain harvesting) has enabled northern farmers to capture and divert floodwaters from the river system without the need for a licence or metre for over 30 years, until it was formally licensed in June 2023 (Baird et al., 2024). Most of the water captured from floodplain harvesting is used for cotton production (Brown et al., 2022), which requires 6–7 million litres of water per hectare (Cotton Australia, 2025). Several of the largest cotton producers in Australia, including Australian Food and Fibre (a subsidiary of the Canadian pension fund PSP Investments) and Peter and Jane Harris, exposed and convicted for water theft in 2020, now control over a quarter of the floodplain water allocations in the Northern Basin (Waldman et al., 2023).

While excessive water extraction has bolstered agriculture and economic activity in the Northern Basin, it has significantly reduced inland flows and disrupted hydrological connectivity along the Lower Darling Barka in south-western NSW (Australian National University, 2025; Baird, 2024). This has left downstream communities grappling with water inequities, including mass fish kills (Australian Academy of Sciences, 2019), deteriorating water quality and related health issues from toxic blue-green algae. In some areas, residents were forced to rely on water tankers for donated supplies and still purchase bottled water from supermarkets (Maloney et al., 2020). Much of the over-extraction of floodplain waters has been licensed and enabled to benefit the corporate irrigation industry, often at the expense of downstream communities and ecosystems, highlighting the critical issue of water justice in the Northern MDB (Baird, 2024; Brown et al., 2022).

In the Southern Basin, particularly southern NSW, Victoria, and South Australia, over-extraction for permanent almond plantings has created inequities for farmers in the water market. Almonds were first planted in Australia on Kangaroo Island in the 1830s, with local orchards appearing in private gardens across South Australia (Almond Board of Australia, 2024). By the 1990s, commercial almond production expanded into the Southern MDB, where the climate conditions and availability of water proved ideal for almond farming. This growth helped establish Australia as one of the world's largest almond exporters, transforming farming practices and water management in the MDB. However, the industry's rapid expansion has raised concerns about water usage and its impacts on other sectors (Davies, 2019).

In the last two decades, the area planted with permanent almond orchards has skyrocketed from 3,500 hectares in 2000 to over 45,000 hectares in 2018 (Williams et al., 2023). This increased to more than 64,000 hectares in 2023 (Almond Board of Australia, 2024). The expansion of the almond industry has been accompanied by an increase in freshwater allocations and the construction of on-farm dams to meet the rising demand for irrigation, often to the detriment of smaller industries like dairy and viticulture, in NSW and Victoria (Kennedy & Mackintosh, 2024; Murray-Darling Basin Authority, 2025). For example, in 2019, general security irrigators along the Upper Murray River in NSW received zero water allocations, because the market demand for irrigation was prioritised for almond orchards downstream in Victoria and South Australia. As a result of this inequitable water distribution, many farmers lost their crops and livelihoods and faced significant mental health challenges, watching freshwater flow past their properties despite paying thousands of dollars annually for their water entitlements (Baird, 2023).

Like cotton production, almonds are among the most water-intensive crops in the MDB, requiring 12.5 million litres of water per hectare, which is far more than other traditional crops grown in the region, such as horticulture, which needs considerably less (Kennedy & Mackintosh, 2024). Following recent plantings, however, the demand for water in the lower Murray River for permanent crops is expected to increase by around 150 gigalitres over the next five years, with almonds accounting for over 70 per cent of that demand, according to the water research firm Aither (cited in Waldman et al., 2023). In years of extreme dryness, intensified by drought and shifting climatic conditions, researchers warn that there won't be enough water to irrigate permanent crops like almonds, let alone meet the needs of other struggling industries, communities and the environment in the MDB (Waldman et al., 2023; Williams et al., 2023).

In water law and policy, water over-extraction is not recognised as illegal behaviour, unlike more traditional offences, such as water theft that involves taking water without a licence. Rather, over-extraction is considered a harmful but legal practice sanctioned by state authorities and bureaucratic powers to support the growth of corporate agribusiness (Baird, 2024). The water market, which facilitates the trading and buying of legal water rights, often incentivises excessive extraction for agricultural and industrial purposes by prioritising freshwater allocations for the highest-yielding crops. This approach favours short-term economic gains for irrigators and corporate agribusiness over the long-term sustainability of fresh water and riverine communities in the MDB.

In the MDB, the growing demand for irrigation to support the expansion of industrial agriculture, particularly cotton and almond farming (Department of Agriculture, Fisheries and Forestry, 2023), has resulted in unsustainable levels of water extraction from the river system, disrupting hydrological connectivity and exacerbating problems like blue-green algae, mass fish deaths and desertification in the MDB (Australian Academy of Sciences, 2019; Brown et al., 2022; Colloff et al., 2024; Murray-Darling Basin Authority, 2025; Waldman et al., 2023; Wentworth Group of Concerned Scientists, 2023). The harm extends beyond impact to the environment with reports of community well-being and cohesion eroding through socioeconomic decay. Many small-scale farmers, local residents (Indigenous and non-Indigenous alike), and other agricultural industries, like dairy, have reported a decline in employment, tourism, population, and recreation within their local communities, and experienced significant mental and physical health challenges due to a lack of access to clean, potable water (Maloney et al., 2020; Sefton et al., 2020). Over-extraction in the MDB, driven by large-scale cotton and almond production for export, deepens water injustice for smaller farmers, local communities, ecological users and First Nations, who all struggle to compete with corporate irrigators for fair access in the water market (Baird, 2024; Environmental Justice Australia, 2023).

4 Expropriation, Exclusion, and Alienation

Since time immemorial, Aboriginal peoples/communities have occupied the MDB as its traditional custodians with over fifty First Nations holding deep spiritual and cultural ties to the landscape, its rivers and ancestors. The riverine plains and valley of the MDB are central to their social life and customary economies (Weir, 2010). However, European colonisation disrupted this connection by introducing foreign plants, animals and an ambition to transform the landscape, exemplified through the notion of *terra* and *aqua nullius*; land and water belonging to no one (Marshall, 2017; see also Harriden, 2021). Settlers sought to establish agrarian societies, claiming their moral ownership over natural resources, especially fresh water, and altering the river system for agriculture, irrigation and mining. Colonial water law granted water rights to white settlers who owned and farmed the land, while Aboriginal people were excluded from land ownership and marginalised by settler-state systems of water governance and administration. This injustice lasted for most of the nineteenth and twentieth centuries (Hartwig et al., 2020).

Following increased momentum from social justice movements to recognise Aboriginal land and water rights throughout the twentieth century, state governments introduced land restitution measures, such as the *Aboriginal Land Rights Act* 1983 (NSW), to compensate Aboriginal peoples for colonial acts of dispossession, but water entitlements were rarely included. Land restitution measures often excluded properties with productive agricultural value, meaning few returned parcels of land came with water rights (Hartwig et al., 2020). Governments entrenched this inequity by strengthening the water rights of those who already held entitlements through official policy and market development processes. Aboriginal water rights were not formally recognised until 1992, following the case of *Mabo v Queensland (No 2)* [1992]

HCA 23; (1992) 175 CLR 1 in the Australian High Court, which prompted the *Native Title Act* 1993 (Cth). While the *Native Title Act* 1993 (Cth) facilitates the recognition of Aboriginal water rights by the state, it generally only covers water rights for personal, social, domestic, and non-consumptive cultural uses. As a result, water rights for agricultural and commercial purposes were excluded from restitution processes (Hartwig et al., 2022).

The moment at which Australian society began to address colonial dispossession and recognised First Nations' land and water rights coincided with significant changes to water access (Hartwig et al., 2020). In 2004, when land and water titles were separated, settler-governments exacerbated the unequal distribution of water inherited from the colonial era by granting existing holders grandfathered water rights and transforming key water resources in the MDB into commodities to establish a formal water market. Consequently, this meant Aboriginal organisations and participants entered the market with very limited water rights (Hartwig et al., 2020). In some cases, Aboriginal participants have benefited from the water market by trading freshwater allocations on the temporary market to generate quick revenue streams. In addition, the revenue from water trading can support greater autonomy for Aboriginal organisations, allowing them to operate outside the constraints of bureaucratic and corporate forces that often limit self-determination (Hartwig et al., 2023).

Regardless of these benefits, however, past, present, and ongoing acts of water expropriation and marginalisation by settler-state authorities, described by Hartwig et al. (2022) as water colonialism, have compounded difficulties faced by Aboriginal people to accumulate wealth and exercise self-determination, while also undermining the recognition and legitimacy of First Nations' water rights in the Australian water market. In the MDB, Aboriginal people hold less than 0.2 per cent ($18.4 million) of all surface water entitlements available for purchase, which are valued at approximately $16.5 billion in the water market (Kennedy et al., 2022). The increasing centralisation of the water market around high yield crops and agricultural expansion, along with the rise of non-landholder participants, creates significant barriers for Aboriginal participants. This forces Aboriginal participants to secure greater capital to compete with powerful market players, or else they may be forced to maximise the economic value of their water rights to avoid financial instability (Hartwig et al., 2023). Between 2009 and 2018, nearly one-fifth of Aboriginal water holdings by volume were lost as organisations were forced to permanently sell their entitlements, often due to impending bankruptcy (Hartwig et al., 2023). While First Nations Australians view water as a holistic resource, the neo-colonial water market pressures them to generate economic returns, often disregarding their own aspirations for water management (Hartwig et al., 2022). In addition, when water and land rights are returned to Aboriginal peoples by the settler-state, the amount of water is minimal, and the land (following decades of transformation and exploitation for agriculture) is often more suitable for grazing rather than irrigation. Consequently, the economic potential of these water rights tends to yield lower returns in the water market (Hartwig et al., 2023).

In 2018, the former Coalition Government allocated $40 million in taxpayer funds to support Aboriginal participants in the water market for both cultural and commercial purposes, to address such widespread inequities (Department of Climate Change, Energy, the Environment and Water [DCCEEW], 2024). However, for reasons unknown, these funds remained unspent for five years. The Federal Labor Government increased this allocation to $100 million as part of the *Restoring Our Rivers Bill* (DCCEEW, 2024). Despite this, according to the ABC (at this time of writing), none of these funds have been made available to Aboriginal participants in the water market (Kennedy, 2024).

The ongoing struggles of Aboriginal peoples to acquire water rights in the MDB are deeply rooted in colonial dispossession. European colonisation disregarded traditional connections to land and water, reallocating these resources to settler agriculture while marginalising Aboriginal communities. Although legal reforms provided limited recognition of Aboriginal water rights, they were confined to non-consumptive uses, excluding agricultural or commercial access. The creation of the water market only deepened these injustices, granting Aboriginal people minimal entitlements and restricting their participation in the market. Despite attempts by the government to address these inequities, Aboriginal communities have yet to receive meaningful support, perpetuating systemic inequalities in water governance and policy that undermine water justice for First Nations Australians.

5 Ecological Water Justice

While environmental justice safeguards fresh water for human benefit, *ecological* water justice is concerned with preserving the ecological integrity of freshwater systems, extending beyond human needs. Deeply rooted within an eco-centric view, ecological water justice acknowledges that fresh water is essential for all life forms and should be valued and protected for both its intrinsic value and practical uses (Kallhoff, 2014).

In 2012, the Murray-Darling Basin Authority (MDBA) launched the Murray-Darling Basin Plan to manage the Basin as an interconnected system, aiming to restore the declining ecological health of the MDB while still supporting farming and agricultural industries (Leblanc et al., 2012). In recent decades, however, freshwater supplies essential for maintaining ecological health in the MDB have steadily declined. A study by the Wentworth Group of Concerned Scientists (2023) revealed that from 1979 to 2022, only 31 per cent of the minimum environmental water flows required to sustain ecosystem health in the MDB were met. Since the Murray-Darling Basin Plan was introduced (which cost over $13 billion in taxpayer funds), only 26 per cent of the necessary environmental water flows have been restored to the river system (Wentworth Group of Concerned Scientists, 2023). There are several factors, including climate change, drought, over-extraction, and artificial constraints, which are threatening the ecological health of the MDB (Colloff et al., 2024).

What makes a river system healthy is a continuous supply and flow of fresh water, which maintains hydrological connectivity between upstream rivers and downstream ecosystems. These connecting flows are essential for good water quality and delivering nutrients throughout the river system, while supporting native flora and fauna. Since European settlement, however, the regulation of fresh water through artificial structures (developed for both agricultural and environmental purposes) has significantly altered and reduced natural flow patterns by 60 per cent (2012 levels) in the MDB. These changes affect water that flows through the Murray Mouth estuary into the ocean in South Australia (Stewardson et al., 2021).

In the Northern Basin, the construction of in-stream dams, weirs, and private on-farm storages has led to more frequent and longer cease-to-flow events, which would have otherwise maintained hydrological connectivity along the Darling Barka. While some rivers and streams in the Northern Basin are naturally perennial, intermittent, or ephemeral, these artificial barriers have impeded hydrological connectivity in the river system, reducing inland flows and hindering native fish migration, floodplain inundation, and the movement of essential nutrients (Stewardson et al., 2021). These effects are also experienced in the Southern Basin, where significant headwater dams and reservoirs have changed the extent, duration, and frequency of flooding events that were once common in the MDB and vital for downstream ecosystems. For instance, the lower Goulburn River, which used to flood twice every five years on average, now only floods once every ten years due to water resource development (Stewardson et al., 2021). In addition, alterations to natural flow patterns in the MDB have affected the ability of surface water systems to connect with and recharge groundwater aquifers across the MDB (Pittock et al., 2023). In essence, artificial structures undermine ecological water justice in the MDB, by damming the river system and restricting its ability to naturally flow in its own right and calls for greater legal recognition of rivers' rights in Australia.

6 Water Justice—Towards an Aquatic Green Criminology

As discussed above, ecological water justice extends to the rights of rivers to flow freely without harm and disruption from human activities (White, 2021). Freshwater bodies should be recognised for their intrinsic value and perceived as independent beings with agency that exists autonomously of human control. For many, water injustice is also a form of political deviance and social harm and the hallmark sign of a failing democracy (Fioret, 2023). A designated and distinct scholarship in aquatic green criminology permits the rich tapestry of existing discourses in environmental crime and green criminology to be devoted to all things water. This seeks to encapsulate existing developments with 'blue crime' that focus on ocean harms and the ways that state-corporate power exploits marine habitats (Bueger & Edmonds, 2020) by extending analysis to all waterscapes.

We argue that green criminology can learn from ocean studies when examining the destruction of both freshwater and marine environments, in creating an aquatic green criminology that examines harm across land and seascapes. When marine justice scholarship harnesses criminological discourses, we observe the emergence and exploration of crimes involving freshwater (whether it be harvested, recycled or reclaimed) and seawater. However, a scholarship not only devoted to water per se, but also the flora and fauna that rely on water habitats for survival. This pushing of new criminological boundaries is most relevant, notably where the world's scientific community has identified devastating planetary indicators requiring renewed and urgent international action (Bulletin of Atomic Scientists, 2025). In doing so, this chapter seeks to extend the parameters of 'southern blue criminology' (Hubschle & Berg, 2024) that focuses on seascapes, to include all aquatic environments, both freshwater and seawater.

The discourse within marine justice is an emerging project that seeks to harness multiple disciplinary knowledges and policies to protect and preserve global ocean environments (Bercht et al., 2021). It was proposed by Longo and Clark (2016: 464) within a framework of 'marine sociology' to articulate humanity's intimate connection and mutual benefit with a healthy ocean environment. It is embedded in the principles of environmental justice and social ecology, notably an emphasis on healthy oceans and the concomitant social, cultural and economic advantages to humanity (Martin et al., 2019). Marine justice is a new multidisciplinary network that coalesces around scientific evidence within marine biology and oceanography to preserve coastal and ocean habitats against climate change and anthropogenic activities (Bercht et al., 2021). It is not a discourse of blame, but one of hope and prosperity that seeks to harness new knowledge to protect aquatic environments.

To date, it has inculcated the expertise of oceanographers, marine scientists and activists to examine 'five conceptual domains – space, time, knowledge, participation in decision-making, and enforcement in which attention to marine environments resonates with and expands environmental framings' (Martin et al., 2019: 234). Similarly, an aquatic green criminology seeks to embrace these frameworks for all waterscapes; recognising that the human right to freshwater and the use and access of seascapes are ever-present to the debates and circumstances of global peace and conflict; as well as the ongoing integrity or despoliation of the natural world. Such burgeoning discourses in aquatic green criminology provide an innovative lens to critique political decision-making about the MDB and other ecologically important waterscapes.

7 Conclusion

This chapter concludes that water justice is essential to the ongoing existence of planetary flora and fauna. The advent of an 'aquatic-green criminology' provides a platform to encourage scholarship and debate about the centrality of waterscapes to all forms of life. There is much to be gained by introducing criminology, and green criminology in particular, to emerging discourses in freshwater and marine justice. Green criminology scholars continue to highlight, examine, and critique the plight of marginalised peoples and communities burdened by water inequity, especially in the MDB. We argue that the concept of 'water justice' should recognise both the intrinsic and instrumental value of water resources across all waterscapes, extending beyond human interests alone. In doing so, such scholarship ensures criminology, and 'aquatic green criminology', remains not only relevant, but at the forefront of human security and planetary protection.

References

Aboriginal Land Rights Act 1983 (NSW).

Almond Board of Australia. (2024). *Almond Insight 2023/24*. ABA. Retrieved January 28, 2025, from https://almondboard.org.au/almond-industry-releases-new-stats/?v=8bcc25c96aa5

Australian Academy of Sciences. (2019, February 18). *Investigation of the Causes of the Mass Fish Kills in the Menindee Region NSW over the Summer of 2018–2019*. AAS. Retrieved December 15, 2025, from https://www.science.org.au/files/userfiles/support/reports-and-plans/2019/academy-science-report-mass-fish-kills-digital.pdf

Australian Competition and Consumer Commission. (2021). *Murray-Darling Basin Water Markets Inquiry: Final Report*. ACCC. Retrieved January 15, 2025, from https://www.accc.gov.au/about-us/publications/murray-darling-basin-water-markets-inquiry-final-report

Australian National University. (2025). *Darling River's Decline Due to Excessive Water Extractions, Not Only Drought*. ANU. Retrieved April 14, 2026, from https://policybrief.anu.edu.au/darling-river-excessive-water-extractions/

Baird, A. (2023). *Drain the State: Exploring Water Theft in Australia's Murray-Darling Basin* (PhD dissertation). Deakin University.

Baird, A. (2024). Farmers as Allies Towards Ecological Justice: Lessons from Water Markets, Colonialism and Theft in Australia's Murray-Darling Basin. In M. Lobo, E. Mayes, & L. Bedford (Eds.), *Planetary Justice* (pp. 39–53). Bristol University Press.

Baird, A., Bedford, L., Walters, R. & White, R. (2024). *Exploring Water Theft: Discussion Paper, ARC Water Theft Project*. Retrieved January 10, 2025, from https://bpb-ap-se2.wpmucdn.com/blog.une.edu.au/dist/3/1351/files/2024/09/MKT0001675-ARC-Water-Theft-Paper-2_digital.pdf

Bercht, A. L., Hein, J., & Klepp, S. (2021). Introduction to the Special Issue "Climate and Marine Justice–Debates and Critical Perspectives". *Geographica Helvetica, 76*(3), 305–314. https://doi.org/10.5194/gh-76-305-2021

Boelens, R. A., Vos, J. M. C., & Perreault, T. (2018). Introduction: The Multiple Challenges and Layers of Water Justice Struggles. In R. Boelens, T. Perreault, & J. Vos (Eds.), *Water Justice* (pp. 1–32). Cambridge University Press.

Brown, P., Colloff, M. J., Slattery, M., Johnson, W., & Guarino, F. (2022). An Unsustainable Level of Take: On-Farm Storages and Floodplain Water Harvesting in the Northern Murray–Darling Basin, Australia. *Australasian Journal of Water Resources, 26*(1), 43–58. https://doi.org/10.1080/13241583.2022.2042061

Bueger, C., & Edmonds, T. (2020). Blue Crime: Conceptualizing Transnational Organized Crime at Sea. *Marine Policy, 119*, 104067. https://doi.org/10.1016/j.marpol.2020.104067

Bulletin of Atomic Scientists. (2025). *It is Now 89 Seconds to Midnight*. Retrieved May 1, 2025, from https://thebulletin.org/doomsday-clock/

Colloff, M., Pittock, J., Grafton, Q., Wheeler, S., Williams, J., Lanyon, K., Moggridge, B., Kingsford, R., & Bino, G. (2024). Bargaining With Nature Over Water Reforms: The First Independent Assessment of the State of the Murray-Darling Basin Raises Serious Concern. *Global Water Forum*. Retrieved May 20, 2025, from https://www.globalwaterforum.org/2024/12/19/bargaining-with-nature-over-water-reforms-the-first-independent-assessment-of-the-state-of-the-murray-darling-basin-raises-serious-concern/

Cotton Australia. (2025). Cotton's Water Use. *Cotton Australia*. Retrieved January 29, 2025, from https://cottonaustralia.com.au/cottons-water-use#:~:text=Australia%E2%80%99s%20cotton%20industry%20is%20one%20of%20the%20most

Davies, A. (2019, May 26). Tough Nut to Crack: The Almond Boom and its Drain on the Murray-Darling. *The Guardian*. Retrieved January 28, 2025, from https://www.theguardian.com/australia-news/2019/may/26/tough-nut-to-crack-the-almond-boom-and-its-drain-on-the-murray-darling

De Garis, S. A. (2013). The Cotton Industry in Australia: An Analysis. Proceedings of the 19th Annual Pacific Rim Real Estate Conference. 13–16 January 2013, Melbourne, Australia.

Department of Agriculture, Fisheries and Forestry. (2023). *Snapshot of Australian Water Markets: Water Demand*. ABARES. Retrieved April 14, 2026, from https://www.agriculture.gov.au/abares/products/insights/snapshot-of-australian-water-markets#water-market-prices-are-driven-by-supply-and-demand

Department of Climate Change, Energy, the Environment and Water. (2024). *Delivering First Nations Water Ownership in the Basin*. DCCEEW. Retrieved January 20, 2025, from https://minister.dcceew.gov.au/plibersek/media-releases/delivering-first-nations-water-ownership-basin

Environmental Justice Australia. (2023). *Progress or Watered Down Laws?* EJA. Retrieved April 14, 2026, from https://envirojustice.org.au/progess-or-watered-down-laws-murray-darling-basin/

Fioret, C. (2023). Water Justice and Socioenvironmental Justice. *Ethics, Policy and Environment, 26*(3), 406–421. https://doi.org/10.1080/21550085.2022.2090211

Global Water Forum. (2025). The Future Water Crisis: Are We Prepared for the Best or Worst Case? Retrieved June 5, 2025, from https://www.globalwaterforum.org/2025/02/13/the-future-water-crisis-are-we-prepared-for-the-worst-or-best-case/

Grafton, Q. (2020, April 14). While Towns Run Dry, Cotton Extracts 5 Sydney Harbours' Worth of Murray-Darling Basin Water a Year. It's Time to Reset the Balance. *The Conversation*. Retrieved April 14, 2026, from https://theconversation.com/while-towns-run-dry-cotton-extracts-5-sydney-harbours-worth-of-murray-darling-water-a-year-its-time-to-reset-the-balance-133342

Harriden, K. (2021, April 14). Resisting Aqua Nullius, Indigenous River Management in Contemporary Australia. *Global Water Forum*. Retrieved May 2, 2025, from https://www.globalwaterforum.org/2021/04/14/resisting-aqua-nullius-indigenous-river-management-in-contemporary-australia/

Hartwig, L. D., Jackson, S., & Osborne, N. (2020). Trends in Aboriginal Water Ownership in New South Wales, Australia: The Continuities Between Colonial and Neoliberal Forms of Dispossession. *Land Use Policy, 99*, 104869. https://doi.org/10.1016/j.landusepol.2020.104869

Hartwig, L. D., Jackson, S., Markham, F., & Osborne, N. (2022). Water Colonialism and Indigenous Water Justice in South-Eastern Australia. *International Journal of Water Resources Development, 38*(1), 30–63. https://doi.org/10.1080/07900627.2020.1868980

Hartwig, L. D., Jackson, S., Smart, J. C., & Osborne, N. (2023). Water Trading by Aboriginal Organisations in NSW, Australia. *Journal of Rural Studies, 100*, 102997. https://doi.org/10.1016/j.jrurstud.2023.03.005

Hubschle, A., & Berg, J. (2024). Southern Blue Criminology: Rethinking Ocean Harmscapes in a Global Context. *Frontiers in Conservation Science, 5*, 1422829. https://doi.org/10.3389/fcosc.2024.1422829

Johnson, H., South, N., & Walters, R. (2016). The Commodification and Exploitation of Fresh Water: Property, Human Rights and Green Criminology. *International Journal of Law, Crime and Justice, 44*, 146–162. https://doi.org/10.1016/j.ijlcj.2015.07.003

Kallhoff, A. (2014). Water Justice: A Multilayer Term and Its Role in Cooperation. *Analyse & Kritik, 36*(2), 367–382. https://doi.org/10.1515/auk-2014-0209

Kennedy, E. (2024, August 4). $100m Indigenous Murray-Darling Basin Water Fund Faces Devaluation Amid Rising Prices, *ABC News*. Retrieved February 10, 2025, from https://www.abc.net.au/news/rural/2024-08-04/murray-darling-basin-aboriginal-water-concerns-government/104155452

Kennedy, E., Kennedy, B. & Chandrashekeran, S. (2022, March 20). Terra Nullius Has Been Overturned. Now We Must Reverse Aqua Nullius and Return Water Rights to First Nations People. *The Conversation*. Retrieved May 20, 2025, from https://theconversation.com/terra-nullius-has-been-overturned-now-we-must-reverse-aqua-nullius-and-return-water-rights-to-first-nations-people-180037

Kennedy, E. & Mackintosh, A. (2024, February 4). Water-Intensive Almonds Now the Largest Crop Grown in the Lower Murray Amid Call for Moratorium. *ABC News*. Retrieved January 28, 2025, from https://www.abc.net.au/news/rural/2024-02-04/almond-industry-expanding-in-australia-high-water-usage/103388210

Leblanc, M., Tweed, S., Van Dijk, A., & Timbal, B. (2012). A Review of Historic and Future Hydrological Changes in the Murray-Darling Basin. *Global and Planetary Change, 80*, 226–246. https://doi.org/10.1016/j.gloplacha.2011.10.012

Longo, S., & Clark, B. (2016). An Ocean of Troubles: Advancing Marine Sociology. *Social Problems, 63*(4), 463–479. https://doi.org/10.1093/socpro/spw023

Mabo v Queensland (No 2) [1992] HCA 23; (1992) 175 CLR 1.

Maloney, M., Boehringer, G., MacCarrick, G., Satija, M., Graham, M., & Williams, R. (2020). *The 2019 Inquiry into the Health of the Barka/Darling River and Menindee Lakes*. Australian Peoples' Tribunal for Community and Nature's Rights, Retrieved January 30, 2025, from https://tribunal.org.au/wp-content/uploads/2020/10/2019CitizensInquiry_BarkaDarlingMenindee-1.pdf

Marshall, V. (2017). Overturning Aqua Nullius: Pathways to National Law Reform. In R. Levy, M. O'Brien, S. Rice, P. Ridge, & M. Thornton (Eds.), *New Directions for Law in Australia* (pp. 221–230). Australian National University Press.

Martin, J., Gray, S., Aceves-Bueno, E., Alagona, P., Elwell, T., Garcia, A., Horton, Z., Lopez-Carr, D., Marter-Kenyon, J., Marie Miller, K., Severen, C., Shewry, T., & Twohey, B. (2019). What is Marine Justice? *Journal of Environmental Studies and Sciences, 9*, 234–243. https://doi.org/10.1007/s13412-019-00545-0

McClanahan, B. (2014). Green and Grey: Water Justice, Criminalization, and Resistance. *Critical Criminology, 22*, 403–418. https://doi.org/10.1007/s10612-014-9241-8

Michel, D. (2023). The Evolving and Incompletely Realised Human Right to Water. *Centre for Strategic and International Studies*. Retrieved March

20, 2025, from https://www.csis.org/analysis/evolving-and-incompletely-realized-human-right-water

Murray-Darling Basin Authority. (2025, April 1). *Irrigation Trends–Choosing What to Grow*. MDBA. Retrieved April 14, 2026, from https://www.mdba.gov.au/water-use/trends-use/irrigation-trends-choosing-what-grow

Native Title Act 1993 (Cth).

O'Donnell, E., Clark, C., & Killean, R. (2024). Rights and Relationality: A Review of the Role of Law in the Human/Water Relationship. *Water Alternatives, 17*(2), 207–238. https://www.water-alternatives.org/index.php/alldoc/articles/volume-17/v17issue2/749-a17-2-7/file

Pittock, J., Corbett, S., Colloff, M. J., Wyrwoll, P., Alexandra, J., Beavis, S., & Williams, J. (2023). A Review of the Risks to Shared Water Resources in the Murray–Darling Basin. *Australasian Journal of Water Resources, 27*(1), 1–17. https://doi.org/10.1080/13241583.2023.2190493

Sefton, R., Peterson, D., Woods, R., Kassebaum, A., McKenzie, D., Simpson, B., & Ramsay, M. (2020). Final Report: Independent Assessment of Social and Economic Conditions in the Murray–Darling Basin, Panel for Independent Assessment of Social and Economic Conditions in the Murray-Darling Basin. Murray-Darling Basin Authority, Retrieved February 15, 2025, from https://www.dcceew.gov.au/sites/default/files/documents/panel-report.pdf

Stewardson, M. J., Walker, G., & Coleman, M. (2021). Hydrology of the Murray-Darling Basin. In B. T. Hart, N. R. Bond, N. Bryon, C. A. Pollino, & M. J. Stewardson (Eds.), *Murray-Darling Basin, Australia: Its Future Management* (pp. 47–73). Elsevier.

Sultana, F. (2018). Water Justice: Why it Matters and How to Achieve It. *Water International, 43*(4), 483–493. https://doi.org/10.1080/02508060.2018.1458272

UN Global Goals. (2025). *Clean Water and Sanitation*. UN. Retrieved March 21, 2025, from https://www.globalgoals.org/goals/6-clean-water-and-sanitation/

UNESCO. (2024). *World Water Development Report 2024: Executive Summary*. UNESCO. Retrieved March 20, 2025, from https://unesdoc.unesco.org/ark:/48223/pf0000388950

United Nations. (2010). *The Right to Water*. UN. Retrieved March 21, 2025, from https://www.ohchr.org/sites/default/files/2021-09/FactSheet35en.pdf

Waldman, P., Rangarajan, S., Whitely, A., & Gross, A. (2023, December 27). *The Water Trade Is Booming—And Sucking Australia Dry*. Bloomberg Press. Retrieved January 30, 2025, from https://www.bloomberg.com/graphics/2023-australia-water-trade-drought/?srnd=phx-green-water-grab

Water Justice Hub. (2024). *About the Hub*. WJH, Retrieved March 24, 2025, from https://www.waterjusticehub.org/about/the-hub/

Water Justice Hub. (2025). *Water Justice Day*. WJH, Retrieved March 24, 2025, from https://www.waterjusticehub.org/water-justice-day

Weir, J. K. (2010). Cultural Flows in Murray River Country. *Australian Humanities Review, 48*, 131–142.

Wentworth Group of Concerned Scientists. (2023). *Are Murray-Darling Basin Rivers Getting the Water They Need to Stay Healthy*. Retrieved February 15, 2025, from https://wentworthgroup.org/wp-content/uploads/2023/09/MDB-EWR-Report-8-Sept-2023-1.pdf

Wheeler, S., Loch, A., Zuo, A., & Bjornlund, H. (2014). Reviewing the Adoption and Impact of Water Markets in the Murray–Darling Basin, Australia. *Journal of Hydrology, 518*, 28–41. https://doi.org/10.1016/j.jhydrol.2013.09.019

White, R. (2021). *Theorising Green Criminology. Selected Essays*. Routledge.

Williams, J., Colloff, M. J., Grafton, R. Q., Khan, S., Paydar, Z., & Wyrwoll, P. (2023). The Three-Infrastructures Framework and Water Risks in the Murray-Darling Basin, Australia. *Australasian Journal of Water Resources, 27*(1), 47–67. https://doi.org/10.1080/13241583.2022.2151106

World Atlas. (2018). How Much of the Earth is Water? World Atlas. Retrieved March 20, 2025, from https://www.worldatlas.com/articles/how-much-of-the-earth-is-water.html#:~:text=Approximately%2071%25%20of%20the%20earth%27s%20surface%20is%20covered

World Economic Forum. (2025). What is World Water Day? WEF. Retrieved March 20, 2025, from https://www.weforum.org/stories/2025/03/world-water-day-march-2025/

Zwarteveen, M. Z., & Boelens, R. (2017). Defining, Researching and Struggling for Water Justice: Some Conceptual Building Blocks for Research and Action. In R. Boelens, B. Crow, J. Hoogesteger, F. E. Lu, E. Swyngedouw, & J. Vos (Eds.), *Hydrosocial Territories and Water Equity* (pp. 8–23). Routledge.

4

'Just Care' for the More-Than-Human World: An Ecofeminist Reimagining of Justice in Green Criminology

Cassie A. Pedersen and Stephen R. Burrell

1 Introduction

Green criminologists have made vital contributions in broadening understanding of 'crime' to encompass harms against humans, nonhuman animals, and ecosystems—accounting for devastating human (in)actions which may not violate any laws. Green criminology has thus expanded the notion of 'justice' to incorporate the more-than-human world (Nurse, 2015), forging more inclusive approaches that move beyond legalistic and anthropocentric frameworks. Despite this important work in expanding the concept of justice, green criminology often proceeds

C. A. Pedersen (✉)
Federation University, Berwick, VIC, Australia
e-mail: c.pedersen@federation.edu.au

S. R. Burrell
University of Melbourne, Melbourne, VIC, Australia
e-mail: stephen.burrell@unimelb.edu.au

R. Killean et al. (eds.), *Conceptualising Justice*, Palgrave Studies in Green Criminology, https://doi.org/10.1007/978-3-032-19791-7_4

without questioning whether the concept itself—shaped by, and entangled with, problematic legacies of Western Enlightenment thought—should remain intact, be discarded, or fundamentally reimagined.

Green criminology is often divided into three distinct 'justice' perspectives, based on the influential work of White (2013; see also Chapter 2). 'Environmental justice' focuses on human wellbeing and equitable access to healthy environments. This sees 'environmental rights' as an extension of human rights and social justice, aiming to enhance the quality of human life. Grounded in anthropocentrism, it implies human distinctiveness and superiority over nature, valorising human-centred perceptions of value and use. 'Ecological justice' focuses on the wellbeing of the biosphere, specific ecosystems, and the interconnected plants and animals inhabiting them, arguing that the environment has its own rights and should be preserved for its own sake. This ecocentric framework does not prioritise a specific facet of ecosystems but sees humans as simply one component within them, suggesting ecological citizenship which extends the moral community to the more-than-human world. 'Species justice' focuses on the wellbeing of nonhuman animals as individuals and species, based on the belief that they have an intrinsic right not to suffer abuse. This is rooted in biocentrism, viewing humans as simply one animal species among others with the same moral worth, emphasising animal rights as *sui generis*.

White (2013) discusses tensions between these approaches, in viewing nonhuman animals and ecosystems instrumentally, prioritising them over one another, or ignoring community interests and inequalities among humans. He contends that they should be placed within an overarching framework of 'eco-justice', which weighs the nature and extent of harm in specific situations, guided by how these different injustices fit together rather than making either/or choices. Prominent green criminologists Lynch et al. (2019: 45) similarly call for an eco-justice approach, 'built on the recognition that there are scientific boundaries to the living world system, and these boundaries become the principles from which a concept of justice can be derived', with human activities that damage the planetary boundaries of 'Gaia' representing unjust acts that threaten the basis of life on Earth. They advocate for examining the *political economic* causes of ecological injustice, in terms of the structural arrangement of the capitalist world system and global and local treadmills of production, as they impede planetary sustainability for human and nonhuman species.

These frameworks reflect important theoretical foundations established within green criminology. Nevertheless, we agree with scholars including Davies (2023), Hunnicutt (2020), and Sollund (2020) that ecofeminism remains a crucial perspective thus far under-utilised within the field. Green criminology has clear overlaps with ecofeminism: their shared efforts to recalibrate human relations with the more-than-human world; questioning of anthropocentric ways of thinking; attention to the marginalised voices of victims of harm; emphasis on humans' distinct ecological responsibility; and contextualised and action-oriented approaches. However, we believe ecofeminism can take green criminological thinking forward in important ways—for example, through addressing patriarchal and colonialist power relations. Whilst green criminological critiques have often emphasised the role of industrial capitalism in driving the destruction of the more-than-human world, capitalism is inseparable from patriarchy and white supremacy in the logics and practices of extractivism.

Ecofeminist and decolonial scholars have charted how the anthropocentrism at the heart of beliefs in human superiority and entitlement to endlessly extract from the Earth is rooted in patriarchal Western ontologies which underpin capitalist and colonialist expansion (Chemhuru, 2018; Plumwood, 1993). At the top of these rationalist, 'scientific' hierarchies is a *specific group* of humans: white, upper-class men, constructed as the embodiments of civilisation and 'culture'. Women, people of colour, and working-class people are constructed as 'less than' precisely because they are associated with 'nature' and thus 'primitiveness'. Ecofeminism therefore enables us to develop genuinely multi-dimensional analyses of the oppression of nature, by illustrating that these different systems of power are not just intersecting, but mutually foundational (Ko, 2019).

Another vital contribution ecofeminism makes to green criminology is in highlighting the origins of contemporary Western notions of 'justice' in the same liberal Enlightenment thought as these anthropocentric ideologies. Justice is a widely used concept, with multiple interpretations and meanings. It is often applied with laudable goals to address the unfair and unequal treatment of marginalised groups, or to advance alternative responses to harm beyond criminal 'justice' systems, such as restorative justice. However, the concept tends to be rooted in liberal notions of the social contract, based upon maximising individual liberty and a Kantian view of people as free, equal, morally autonomous, rational agents (Schlosberg, 2007). This perspective has been critiqued from multiple directions, including by advocates of (alternative definitions of) justice themselves; however, it brings into question whether justice always provides the most useful ethical framework for green criminologists. (Eco)feminist scholars have illustrated that these ethical frameworks are not gender-neutral; justice perspectives have taken prominence because they are rooted in a *masculinised* way of thinking, juxtaposed with a feminist emphasis on care, situatedness, and relationality.

An ethic of justice is characterised by an impersonal focus on abstract, universal principles and rules, upholding equality before the law, fairness, and individual rights as core values (Luke, 2007). In contrast, an ethic of care is attentive to context, emphasising relationships and responsibility. (Eco)feminists have argued that the 'justice' approach derives from a masculine ontology in which the self is seen as a separate, independent, and autonomous entity (Plumwood, 1991). In contrast, the care approach embodies a feminist relational ontology, where the self is experienced as interconnected with others. Whereas the justice approach has taken precedence, the care approach has historically been marginalised by Western patriarchal modes of thought—as is the case with more relational, First Nations ontologies. We argue that care demands greater attention, not least from green criminologists, for whom it offers a different way of conceiving of our relationship with our ecologies (Hunnicutt, 2020). Nevertheless, rather than upholding a false dichotomy between 'care' and 'justice', in this chapter we seek to deconstruct this patriarchal dualism by developing an ecofeminist 'just care' concept, which grounds the pursuit of justice in one's embodied experiences and caring relationships with the more-than-human world. The chapter concludes by applying this framework to a case study of salmon farming in Tasmania, deliberately mobilising an example involving animals markedly different to humans to illustrate that 'just care' is not limited to similar species and is a powerful ethical force promoting justice across difference.

2 'Justice' and 'Care' in (Eco)Feminist Thought

Carol Gilligan (1982/1993) juxtaposes the concepts of 'justice' and 'care' through a critique of Kohlberg's theory of moral development. Kohlberg characterises moral reasoning as progressing through a series of stages: from self-interest, to conformity with social norms and, ultimately, to reasoning based on the abstract, universal principle of justice. Gilligan argues that Kohlberg's framework exhibits a masculine bias that privileges patriarchal (and, we would add, colonialist) ways of being and knowing. In response, Gilligan attunes herself to a 'different voice'—one grounded in relationships, responsibility, and care—which has been traditionally silenced and ignored.

The 'justice' and 'care' perspectives are contrasted through the moral problem-solving of two eleven-year-old children (Gilligan, 1982/1993). Jake and Amy were presented with a moral dilemma, in which a man named Heinz must decide whether to steal an unaffordable life-saving drug from a pharmacist. Jake likens the dilemma to 'a math problem with humans' (Gilligan, 1982/1993: 26), arriving at its resolution through logical inferences. Jake formulates a conflict between life and property, concluding that Heinz should steal the drug because the preservation of life outweighs the right to own property. While recognising theft as legally wrong, Jake upholds that the pharmacist's withholding of the life-saving drug is morally unjustifiable, maintaining that the law is not absolute, and instead appealing to a universal and principled notion of justice (Gilligan, 1982/1993).

Amy prioritises the relationship between Heinz and his wife over appeals to rights and justice. She reasons that while stealing the drug might initially save his wife, Heinz could end up imprisoned, leaving him unable to care for her or obtain further treatment if her condition worsens. Amy thus views the situation through a lens of interconnected relationships, where the moral issue arises not from abstract rules but from the pharmacist's failure to respond compassionately to the wife's needs (Gilligan, 1982/1993). Rather than suggesting Heinz steal the drug, Amy proposes that if he were to convey the urgency of his wife's condition to the pharmacist, then surely, they could reach an agreement—for example, the pharmacist might agree to give him the medication right away, with the understanding that Heinz would pay him back later. Hence, where Jake's perspective aligns with the justice approach, Amy's reasoning reflects an ethic of care—her 'belief in the restorative activity of care, leads her to see the actors in the dilemma arrayed not as opponents in a contest of rights but as members of a network of relationships on whose continuation they all depend' (Gilligan, 1982/1993: 30).

The contrast between Jake and Amy's moral problem-solving is attributed to their different conceptions of self (Gilligan, 1982/1993). Jake represents 'a self defined through separation, … a self measured against an abstract ideal of perfection', while Amy embodies 'a self delineated through connection … a self assessed through particular activities of care' (Gilligan, 1982/1993: 35). These distinct orientations extend beyond interpersonal ethics to inform how humans relate to the more-than-human world. Jake's justice-based reasoning, grounded in individuation and abstraction, mirrors the dominant Western patriarchal worldview that positions 'man' as separate from, and superior to, nature. In contrast, Amy's care-based approach, rooted in relationality, interdependence, and responsibility, positions humans in a web of relations with the more-than-human world. This relational emphasis resonates with First Nations worldviews, which we explore later in this chapter.

While it is tempting to posit 'justice' and 'care' as mutually exclusive, we argue that this dichotomy is neither necessary nor productive. Nevertheless, literature on 'rights' versus 'care' in animal ethics often reinforces this division. Regan (1983)—a staunch proponent of non-sentimentalist reasoning—argues from a deontological perspective that animals possess inherent rights as 'subjects-of-a-life'. He criticises care ethics for being overly emotional and particularistic, warning it risks only being applicable to one's immediate circle of relationships. Likewise, Singer (1975/2015) emphasises rationality over sentimentality, advocating from a utilitarian perspective for the equal consideration of animals' interests.

Within green criminology, Beirne (2007) acknowledges the care ethic perspective and feminist critiques of justice-based frameworks, but ultimately advocates for a greater emphasis on animal rights to address the multitude of harms perpetrated against nonhuman animals. Luke (2007), conversely, argues that animal liberation is better framed through care than justice and rights, emphasising the cultivation of emotional capacities such as sympathy, kindness, and compassion.

We agree that care ethics foster reciprocal, non-exploitative, and mutually enriching relationships between humans and the more-than-human world; however, we maintain that care and justice need not be understood as mutually exclusive. We further challenge the contention that care is parochial and that emotionality should yield to detached rationality in the pursuit of justice. This either/or framing—where one must choose between reason *or* emotion, justice *or* care—reflects a reductive, patriarchal mode of thinking that can be effectively challenged through an ecofeminist lens. The concept of 'just care', which we develop in this chapter, rejects the characterisation of justice and care as oppositional, instead positing care as the foundation of a broader and more inclusive approach to justice.

First, however, it is important to refute any charges that care ethics legitimises gender essentialism, coloniality, and other oppressive frameworks. Indeed, an ethic of care has been accused of promoting a Nietzschean 'slave morality' (Puka, 1993/2016)—risking celebrating values like nurturing and self-subordination which are outcomes of patriarchal, capitalist, and colonial systems that assign caregiving roles to women and minoritised groups based on rigid divisions of gender, race and class. Gilligan (1986: 327), however, espouses a *critical* view of care that calls into question the traditional equation of care with self-sacrifice. '[T]he care perspective ... is neither biologically determined nor unique to women'—including women of colour, who are disproportionately assigned care work. Gilligan's protégés, such as Chu (2014), demonstrate that boys also express deep emotional and relational capabilities in childhood, before being fully inculcated into masculine norms. More recently, Gilligan (2023) frames care as a 'human voice' of resistance to patriarchy. Therefore, it is crucial to distinguish between a *feminised* notion of care and a *feminist* ethic of care. The concept of 'just care' moves beyond the masculine/feminine binary and actively challenges patriarchal constructions of gender. Rather than framing care as the loss of self in relationships or the relegation of women to caregiving roles, a feminist ethic of care rejects the dichotomy of feminine dependence and masculine independence, and instead emphasises *inter*dependence and the preservation of the self *within* relationships.

3 From 'Either/Or' to 'Both/And': Formulating 'Just Care'

Ecofeminists challenge a network of hierarchical dualisms structuring Western thought (Plumwood, 1991, 1993; Warren, 1990), including masculine/feminine, culture/nature, human/nature, human/animal, mind/body, and reason/emotion. These dualisms privilege one side over the other—specifically, masculine over feminine, culture over nature, human over nature, human over animal, mind over body, and reason over emotion (Burrell & Pedersen, 2024). They comprise an interrelated network, where masculinity, culture, human, mind, and reason are systematically linked and positioned as dominant, while femininity, nature, animality, the body, and emotion are granted an associative and subordinated status (Plumwood, 1991). This logic legitimises domination by framing those associated with nature as objects to be controlled, managed, and exploited by those aligned with culture and reason (namely, white Western 'man') (Burrell & Pedersen, 2024).

Such dualisms are not only sexist, but racist, classist, ableist, and speciesist. Through the process of othering, dualisms position minoritised groups on the inferior, subordinated side, aligning those who are racialised, animalised, colonised, gendered, disabled, and economically marginalised with 'nature'—casting them as less rational, less civilised, and ultimately, less human (Ko, 2020; Plumwood, 1993; Taylor, 2017). For instance, First Nations people in Australia were denied recognition as fully human under the doctrine of *terra nullius*, which declared the land uninhabited, justifying its occupation by British invaders (Samson, 2008). In Western philosophical and scientific traditions, women have long been portrayed as more emotional, irrational, and governed by the body, in contrast to men who are associated with reason, mind, and culture—justifying their exclusion from public, political, and intellectual life (Plumwood, 1993). Through transatlantic slavery, Black people have been depicted as physically strong but intellectually inferior, closer to animals and nature, to legitimise their enslavement and exclusion from full humanity (Strings, 2019). Working-class people are portrayed as ruled by bodily needs and instincts, lacking the refinement and rationality attributed to the upper-class (Hribal, 2007). Disabled people have been medicalised and institutionalised, perceived as lacking reason and autonomy, and thus positioned outside the bounds of normative humanity (Taylor, 2017). These examples are not merely historical; the hierarchical logic of dualisms continues to perpetuate the marginalisation and dehumanisation of subordinated groups to this day.

It is crucial to note that such dualisms are largely specific—although not exclusive—to Western colonialist thought. For example, First Nations perspectives in Australia emphasise a deep interconnection between humans and nature through the concept of Country (Cameron, 2020). As Deborah Bird Rose (1996: 8) explains:

> Country is multi-dimensional—it consists of people, animals, plants, Dreamings; underground, earth, soils, minerals and waters, air ... People talk about country in the same way that they would talk about a person: they speak to country, sing to country, visit country, worry about country, feel sorry for country, and long for country.

This challenges the anthropocentric foundations of dominant Western thought by recognising the natural world as a living, sentient presence embedded in social and spiritual life (Cameron, 2020). Despite ongoing efforts to erase and marginalise these perspectives, they are enduring—resisting the colonial logic that continues to shape global systems and institutions. Ecofeminists expose how these dualisms continue to assume hegemonic status. Rather than reinforcing 'either/or' distinctions, as is the case with hierarchical dualisms, ecofeminists advocate for a 'both/and' perspective—one that embraces complexity, interdependence, and the entanglement of human and more-than-human life.

3.1 A Third Way: Nature-Self Relationality

One proposed strategy for moving beyond dualistic logic involves affirming the subordinated side of the dualism. In the case of masculine/feminine and culture/nature, this means revaluing the feminine and reclaiming the association between women and nature. However, this approach—which Plumwood (1993: 31) terms an 'uncritical reversal'—ultimately preserves the structure of the dualism, reproducing biological essentialism. An alternative strategy involves severing the association between femininity and nature by repositioning women alongside men on the side of culture. However, incorporating women into the dominant side of the dualism fails to challenge the underlying devaluation of nature and elevation of culture. Hence, Plumwood (1993: 36) advocates for 'a third way which does not force women into the choice of uncritical participation in a masculine-biased and dualised construction of culture or into accepting an old and oppressive identity as "earth mothers": outside of culture, opposed to culture, not fully human'. The critical ecofeminist approach rejects hierarchical dualisms altogether, insisting that women *and* men both belong to, and are shaped by, nature *and* culture. This leads to a rethinking of what counts as human—recognising the material, emotional, and other qualities historically associated with femininity and nature as no less human than rationality, culture, and other traits traditionally aligned with masculinity.

It is worth emphasising that positing humans as a part of nature is not the same as placing them in a state of undifferentiated nature. This marks a key point of divergence between Plumwood's (1991) position and certain expressions of ecocentrism found within deep ecology. Not all deep ecologists are guilty of this conflation; indeed, White (2013) and Brisman and South (2018) emphasise that ecocentrism recognises that humans possess a distinct ability to shape systems of production and subsequently bear responsibility to operate within planetary limits. Nevertheless, Plumwood (1991) critiques the tendency to define the central problem in human–nature relations as one of fundamental separation, and the identification of humans with nature as its resolution. This perspective, which she refers to as the 'indistinguishability account', dissolves the boundaries between humans and nature, positing humans as 'just one strand in the biotic web, not the source and ground of all value' (Plumwood, 1991: 12). Unifying the sides of the human/nature dualism erases all meaningful distinctions between humans and other aspects of nature, espousing a reductive framework that homogenises their identities and obliterates difference. In contrast, Plumwood's (1991) approach provides a novel means for recognising both the continuity *and* distinctiveness of humans and nature—acknowledging the embeddedness of humans in the natural world while also respecting the specificity and autonomy of nature. This resists collapsing humans entirely into nature (an approach associated with some strands of ecocentrism) or elevating them above it (as is the case with anthropocentrism), instead positioning humans as simultaneously part of, and distinct from, nature. This represents a rejection of the either/or logic underlying patriarchal dualisms in favour of a feminist both/and logic.

Plumwood (1991, 1993) advocates for a relational perspective that emphasises interconnectedness without collapsing difference. Recognising that people's interests are shaped through relationships does not mean they are rendered indistinct or subsumed into a collective whole. This is at the core of what Plumwood (1991: 20) terms '*self-in-relationship*'—a concept which 'bypasses both masculine "separation" and traditional feminine "merger" accounts of the self' and 'provide[s] an appropriate foundation for an ethic of connectedness and caring for others'. In view of this, an ethic of care should not be mistaken for a self-sacrificial responsiveness to others that neglects one's own needs or dissolves the self in relationships—tropes that lie at the heart of patriarchal constructions of femininity used to dominate women. Instead, an ethic of care is grounded in a strong and situated sense of self, one that is 'embedded in a network of essential relationships with distinct others' (Plumwood, 1991: 20). This ethic emphasises particularity and attachment—expressed through 'connection to and friendship for particular places, forests, animals, to which one is particularly strongly related or attached, and towards which one has specific and meaningful, not merely abstract, responsibilities of care' (Plumwood, 1991: 21).

3.2 Reimagining Justice in Green Criminology: 'Self-in-Relationship' and 'Just Care'

Ecofeminist insights offer green criminology a valuable and much-needed reimagining of human–nature relationships, which neither replicates nor blends together the dominant ecophilosophical paradigms of anthropocentrism, biocentrism, and ecocentrism. This avoids the trappings of anthropocentrism in that it does not valorise humans over nature, as is the crux of the human/nature dualism. Rather, the concept of 'self-in-relationship' central to a critical ecofeminist care ethic respects the needs and interests of both humans and nature, while acknowledging their interdependence and distinctiveness. This enables one to take social justice issues seriously, including the unequal distribution of natural 'resources' and the disproportionate impact of environmental harm on women and other marginalised communities (Bahar, 2023). Nevertheless, it also shows that humans have disproportionate responsibility to care for nature and act urgently and decisively to stop the destructive actions causing unprecedented environmental degradation. This perspective moves beyond the potential limitations of more extreme biocentric viewpoints, which grant all organisms and entities within the biosphere equal value, yet often overlook the relational dynamics between constituent parts of an ecosystem—indeed, an ecofeminist care ethic is concerned with parts in *relation* to the whole. Similarly, it addresses the shortcomings of more extreme ecocentric viewpoints—specifically the 'indivisibility thesis' as identified by Plumwood (1991)—by resisting the mutual exclusivity between the part versus the whole—rejecting dualistic thinking and bringing the part and whole into dialogue. Perhaps most importantly, ecofeminism identifies the *patriarchal, colonialist* lens through which destructive dualistic thinking operates—revealing that the logic used to justify human domination over nature and nonhuman animals is the same used to oppress racialised, colonised, animalised groups of humans.

We further propose that an ecofeminist care ethic provides green criminologists with a novel way of conceptualising justice. As discussed, (eco)feminists juxtapose two gendered ethical and ontological frameworks: a masculinised 'justice' approach and a feminised 'care' approach.

The critical notion of 'care' is not based on the oppositional framework of justice necessitating an impersonal, separate, and autonomous masculinised subject versus care necessitating a passive, self-sacrificial, and self-abnegating feminised object. Rather, an ethic of care is based on self-in-relationship, where the integrity of the self is preserved through and within connections with others. Nevertheless, we recognise that the term 'care' may inadvertently risk being interpreted through a patriarchal, essentialist lens. Our aim is not to replace the dominant, higher-order side of the dualism (i.e. justice) with the subordinated, lower-order side (i.e. care). To this end, we propose the concept of '*just care*', which moves beyond the reductive framing of 'justice as masculine' and 'care as feminine' and combines the relational, affective qualities of care with the normative, rights-based commitments of justice.

'Just care' involves using one's particular embodied experience as the basis for drawing more general, universal inferences. It entails caring for and about concrete others and using this as a foundation for broader questions of justice. It means direct relationships provide the starting point for one's engagement with and treatment of the more-than-human world. However, it does not stop there. It takes these relationships with particular trees, plants, rocks, mountains, rivers, forests, and animals, and uses these as the basis for more general, principled pursuits of justice. It is about mobilising care in its multiple dimensions: to care *for* and *about* justice for all, including nonhuman animals, ecosystems, and all life within the biosphere—including, but not limited to, humans. It appreciates that all lifeforms necessarily exist in relation to one another, but that relationships should not be subsuming. This means not only challenging the hierarchy between humans and the more-than-human world, but also dismantling the hierarchies between different groups of humans. From an ecofeminist perspective, oppressions and injustices are interconnected; therefore, the liberation of one group cannot come about without the liberation of all (Burrell & Pedersen, 2024). 'Just care' is thus not limited to a parochial concern for beings in one's immediate locality. We maintain that it can mobilise care about bigger picture, more distant issues too, promoting ethical decision-making grounded in shared planetary relationships. This includes caring for and about beings who may seem very different to us as humans, such as fish. To illustrate this, in the final part of the chapter we will focus on the example of salmon farming in Tasmania.

4 The Case Study of Tasmanian Salmon Farming

During the writing of this chapter, the Australian Labor Government fast-tracked amendments to the *Environment Protection and Biodiversity Conservation Act* 1999 (Cth) to remove provisions for the Environment Minister to reconsider salmon farming in Macquarie Harbour, West Tasmania. The decision was justified as safeguarding approximately 120 aquaculture jobs, despite widespread criticism of the industry (Balen, 2025). Commercial salmon farming is highly industrialised. Atlantic salmon are farmed in sea cages, where fish faeces and uneaten food accumulate on the seabed, harming sensitive aquatic habitats (Flanagan, 2021). This depletes oxygen in the already climate-stressed aquatic ecosystem, causing suffocation to the Maugean skate—the world's rarest skate, found only in Macquarie Harbour, known as the 'thylacine of the sea' for dating back to the Cretaceous period—threatening their extinction.

Factory farming also causes immense pain and suffering to salmon. The cramped conditions enable parasites to proliferate, creating disease outbreaks which can spread to other species. Tasmanian farms have reportedly fed salmon antibiotics in attempting to prevent such diseases (Flanagan, 2021). In February 2025, at least 1 million salmon died, dumped at landfill sites and rendering plants, after a disease outbreak in south-east Tasmania. Drone footage found workers pumping and sealing living and dead salmon into tubs together (Morton & Burton, 2025). Declining oxygen levels can also cause stress and mass suffocation to salmon. Research has found as many as one in four salmon in farms showing signs of severe depression, leading to stunted growth and fish floating lifelessly at the ocean's surface (Vindas et al., 2016). Some manage to escape, impacting upon local ecosystems. To deter animals such as seals drawn to pens for food, farms use methods including beanbag bullets and explosive charges, causing some to be blinded, deafened, or killed (Flanagan, 2021).

For green criminologists, this case encapsulates White's (2013) justice frameworks. Environmental justice might focus on the despoliation of the environment for humans living close to salmon farms, the wider impacts of feeding fish antibiotics, and the right of farm workers to sustainable jobs. Ecological justice might highlight the destruction of the ecosystem and its inhabitants by the industry. Species justice might articulate the right of all affected animals, including the skate *and* salmon, to live free from exploitation and extinction. All these considerations are important, and can be incorporated together into an eco-justice perspective. 'Just care' goes further still—urging consideration of all groups impacted by this industry not in abstract impersonal terms, but through a feminist lens that centres context, power, emotion, and relationality, prompting questions about the acceptability of salmon farming's continued existence.

4.1 Applying 'Just Care'

A launchpad for 'just care' is attending to the relationships between human and nonhuman beings and the shared ecologies we inhabit. This applies not only to those living and working on the shores of Tasmania. Many consumers eat the fish killed in these farms. *All* of us can advocate for the harms to be stopped, when we comprehend the damage being done—such as the potential extinction of a fish tens of millions of years old—by the society we are members of, justified by notions of our superiority as a species. It is important to recognise both human distinctiveness *and* closeness vis-a-vis other animals and nature—as selves-in-relationship—not to claim pre-eminence, but to understand our responsibility to care for them, given the unprecedented destruction we are *causing* them.

Such claims for justice should not only be based on a generalised awareness of the industry's impacts on different populations; they can also be mobilised by attention towards the particular, such as the lifeworlds of individual Maugean skate. These ray-like fish have long, pointy, sensitive snouts which they use to hide their quadrangular-shaped bodies in the seabed (Cradle Coast Authority, 2024). The undersides of their snouts are covered with tiny electro-sensory organs which help them find other animals to eat. Behind their eyes they have large respiratory openings called spiracles, which pump oxygenated water into their mouths. This enables them to bury themselves in the sediment and breathe without needing to expose their gills. They have been observed to 'walk' slowly across the seabed, using their pelvic fins akin to a pair of 'legs' (Cradle Coast Authority, 2024).

The concept of 'just care' also invites reflection on the lives of individual salmon within intensive farming—and what life would be like free from it, in their own native habitat (Bossi, 2023). This encourages the centring of lived experiences of salmon, not as abstract units of production, but as sentient beings embedded in relational ecologies. Bossi (2023: 42–43) cites the poignant reflection of Fiona Hamilton, a Trawlwulwuy woman and Tasmanian Aboriginal Heritage Officer:

> When I think about the salmon industry as a First Nations person as a sovereign for my sea country, as a black woman, I can't help but to see things through the salmon's eyes. I imagine myself swimming in places where I am forced to be but have no relationship with the people or other kinship ties to the creatures around me. I am caged. I am fed not for myself or for sustainability, but as a product. When I escape, I am hunted while other creatures who attempt to incur within my pen are brutalised. I am one of many, but I am alone, and I witness the country around me disintegrate and become toxic .... I am alive but I am not living ... I carry my sickness in my body and cannot free myself of it. I can only transmit it .... I am understood in my pen in the same way that black women are understood within colonialism.

Grounded in her positionality as a First Nations woman subjected to racist and sexist oppression, Hamilton bears witness to the interconnected harms of colonialism, ecological degradation, and species injustice. Contemplating the experiences of individual fish stirs emotional connection—often a more powerful motivator for change than detached, rational explanations about the consequences of this industry. First

Nations peoples—whose presence in Tasmania was nearly entirely eradicated—demonstrate that alternative modes of relating are eminently possible.

The lives of humans affected by this industry should also be considered—and there is a substantial protest movement of local people calling for the salmon farming companies operating on the island to leave (Balen, 2025). This is a difficult and unpleasant job for workers, who see few of the profits of this multimillion-dollar business (Flanagan, 2021). It is vital that alternative, non-destructive livelihoods are established for them as part of the transition to a green economy (though it is notable that the same levels of concern expressed for this male-dominated industry are often not expressed for other, more feminised sectors). However, by centring relationships, including with nonhuman beings, we are reminded that humans cannot survive, let alone make profit, if we destroy all other lifeforms in our ecologies, and convert them into mere commodities.

5 Conclusion

This chapter has advocated for deeper integration of ecofeminist thought within green criminology, in critically examining the patriarchal dynamics shaping human relations with the more-than-human world and underpinning the conceptual foundations of justice itself. We argue that justice—often expanded but not always interrogated in green criminological discourse—requires reimagining through a critical ecofeminist lens. To this end, we have formulated the concept of 'just care': a distinct ecophilosophical paradigm that resists the confines of anthropocentrism, biocentrism, and ecocentrism. Rather than centring the human, individual lifeforms, or ecosystems, 'just care' emphasises the relational integrity between them. It offers a novel formulation of justice—distinct from environmental-, species-, ecological-, and eco-justice—assuming the ontology of self-in-relationship as its foundation. 'Just care' recognises interdependence without subsumption, attending to particular beings while extending outward toward a principled, relational form of justice which mobilises care as a generative force for justice across the more-than-human world.

Notions of justice are undoubtedly important within green criminology, and in cases such as Tasmania's salmon farming industry, formal justice mechanisms can provide valuable tools for environmental and animal activists (Nurse, 2015). However, we argue that particular, situated, emotional experiences, relationships, and stories can offer a transformative foundation for recognising how each of us connects on a personal and political level to ecological issues, locally, nationally, and globally. This reveals the web of relationships at play in industries such as salmon farming, the multi-layered ripples of harm being caused, and how they touch each of us. It can help to foster empathy with the various beings affected, including those very different to humans—and mobilise affect which drives decision-making based on care and reducing harm, for all involved. It facilitates an understanding of the historically co-constituted power inequalities at play in this contemporary story, including patriarchy, colonialism, capitalism, and anthropocentrism, and how they influence individual lives and whole communities, with the only people ultimately benefitting being a small number of multinational business owners. 'Just care' does not preclude broad judgements about what justice can and should look like, for the multiple, interrelated groups and ecosystems affected—but invites such judgements to be made not in an abstract, detached way, but through connection, empathy, and emotional investment. Even with all the knowledge and power of twenty-first century Western 'justice' systems, industries such as this remain not just legal but facilitated by governments. This suggests that different, more caring courses of action are urgently needed, if we are to create societies which genuinely enable the sustenance and thriving of life in all its diversity and potential. Green criminology has a vital role to play in shedding light on these stories, and on alternative ways forward. We cannot see ourselves as being detached observers to them—we all have a responsibility to help bring about just care.

References

Bahar, H. I. (2023). The Green Road Project and Women's Green Victimisation in Turkey. In E. Milne, P. Davies, J. Heydon, K. Peggs, & T. Wyatt (Eds.), *Gendering Green Criminology* (pp. 187–204). Bristol University Press.

Balen, C. (2025). Tasmania's Salmon Industry Celebrates Law Amendments Protecting Macquarie Harbour Aquaculture as Environmentalists Promise Fight. ABC News, 27 March. https://www.abc.net.au/news/2025-03-27/tasmania-salmon-farm-industry-environment-laws-greens/105101510

Beirne, P. (2007). Animal Rights, Animal Abuse and Green Criminology. In P. Beirne & N. South (Eds.), *Issues in Green Criminology* (pp. 55–84). Willan.

Bossi, L. (2023). How a Sense of Place may Return the Social License to Operate Concept Back to an Ethics of Responsibility Within a Neoliberal Framework—Tasmanian Salmon Story. *Social Licence and Ethical Practice: Research in Ethical Issues in Organizations, 27*, 25–46. https://doi.org/10.1108/S1529-2096202327

Brisman, A., & South, N. (2018). Green Criminology and Environmental Crimes and Harms. *Sociology Compass, 13*(1), 1–12. https://doi.org/10.1111/soc4.12650

Burrell, S. R., & Pedersen, C. (2024). From Men's Violence to an Ethic of Care: Ecofeminist Contributions to Green Criminology. *Journal of Criminology*, advance online publication 25 October, https://doi.org/10.1177/26338076241293145.

Cameron, L. (2020). "Healthy Country, Healthy People": Aboriginal Embodied Knowledge Systems in Human/Nature Interrelationships. *The International Journal of Ecopsychology, 1*(1), 1–14. https://doi.org/10.55671/2767-1380.1015

Chemhuru, M. (2018). Interpreting Ecofeminist Environmentalism in African Communitarian Philosophy and *Ubuntu*: An Alternative to Anthropocentrism. *Philosophical Papers, 48*(2), 241–264. https://doi.org/10.1080/05568641.2018.1450643

Chu, J. Y. (2014). *When Boys Become Boys: Development, Relationships, and Masculinity*. New York University Press.

Cradle Coast Authority. (2024). Maugean Skate: Zearaja Maugeana. Retrieved May 25, 2025, from https://cradlecoast.com/projects/maugean-skate-zearaja-maugeana/

Davies, P. (2023). Eco-feminism and the Gendering Green Criminology Project. In E. Milne, P. Davies, J. Heydon, K. Peggs, & T. Wyatt (Eds.), *Gendering Green Criminology* (pp. 17–33). Bristol University Press.

Environment Protection and Biodiversity Conservation Act 1999 (Cth).

Flanagan, R. (2021). *Toxic: The Rotting Underbelly of the Tasmanian Salmon Industry*. Penguin.

Gilligan, C. (1982/1993). In a Different Voice: Psychological Theory and Women's Development. Harvard University Press.

Gilligan, C. (1986). Reply by Carol Gilligan. *Signs, 11*(2), 324–333. https://doi.org/10.1086/494226

Gilligan, C. (2023). *In a Human Voice*. Polity Press.

Hribal, J. C. (2007). Animals, Agency, and Class: Writing the History of Animals from Below. *Human Ecology Review, 14*(1), 101–112. http://www.jstor.org/stable/24707647.

Hunnicutt, G. (2020). *Gender Violence in Ecofeminist Perspective: Intersections of Animal Oppression, Patriarchy and Domination of the Earth*. Routledge.

Ko, A. (2019). *Racism as Zoological Witchcraft: A Guide to Getting Out*. Lantern Books.

Ko, S. (2020). By "Human," Everybody Just Means "White". In A. Ko & S. Ko, *Aphro-ism: Essays on Pop Culture, Feminism, and Black Veganism from Two Sisters* (pp. 20–27). Lantern Publishing.

Luke, B. (2007). Justice, Caring, and Animal Liberation. In J. Donovan & C. Adams (Eds.), *The Feminist Care Tradition in Animal Ethics* (pp. 125–152). Columbia University Press.

Lynch, M., Long, M. A., & Stretesky, P. (2019). *Green Criminology and Green Theories of Justice*. Palgrave Macmillan.

Morton, A., & Burton, B. (2025). More than a Million Salmon Dumped After 'unprecedented' Mass Mortality at Tasmanian Fish Farms. *The Guardian*, 8 March. https://www.theguardian.com/australia-news/2025/mar/08/million-salmon-dumped-in-landfill-after-unprecedented-mass-death-of-at-tasmanian-fish-farms-ntwnfb

Nurse, A. (2015). *An Introduction to Green Criminology and Environmental Justice*. Sage.

Plumwood, V. (1991). Nature, Self, and Gender: Feminism, Environmental Philosophy, and the Critique of Rationalism. *Hypatia, 6*(1), 3–27. https://doi.org/10.1111/j.1527-2001.1991.tb00206.x

Plumwood, V. (1993). *Feminism and the Mastery of Nature*. Routledge.

Puka, B. (1993/2016). The Liberation of Caring: A Different Voice for Gilligan's "Different Voice". In M. J. Larrabee (Ed.), An Ethic of Care: Feminist and Interdisciplinary Perspectives (pp. 215–239). Routledge.

Regan, T. (1983). *The Case for Animal Rights*. University of California Press.

Rose, D. B. (1996). *Nourishing Terrains: Australian Aboriginal Views of Landscape and Wilderness*. Australian Heritage Commission.

Samson, C. (2008). The Rule of Terra Nullius and the Importance of International Human Rights for Indigenous Persons. *Essex Human Rights Review, 5*(1), 1–12.

Schlosberg, D. (2007). *Defining Environmental Justice: Theories, Movements, and Nature*. Oxford University Press.

Singer, P. (1975/2015). Animal Liberation: A New Ethics for Our Treatment of Animals. HarperCollins.

Sollund, R. (2020). The Victimisation of Women, Children and Non-human Species Through Trafficking and Trade: Crimes Understood Through an Ecofeminist Perspective. In A. Brisman & N. South (Eds.), *Routledge International Handbook of Green Criminology* (2nd ed., pp. 512–528). Routledge.

Strings, S. (2019). *Fearing the Black Body: The Racial Origins of Fat Phobia*. New York University Press.

Taylor, S. (2017). *Beasts of Burden: Animal and Disability Liberation*. The New Press.

Vindas, M. A., Johansen, I. B., Folkedal, O., Höglund, E., Gorissen, M., Flik, G., Kristiansen, T. S., & Øverli, Ø. (2016). Brain Serotonergic Activation in Growth-Stunted Farmed Salmon: Adaption Versus Pathology. *Royal Society Open Science, 3*(5), 160030. https://doi.org/10.1098/rsos.160030

Warren, K. (1990). The Power and Promise of Ecological Feminism. *Environmental Ethics, 12*(2), 125–146.

White, R. (2013). *Environmental Harm: An Eco-Justice Perspective*. Policy Press.

5

Seeking Eco-Justice by Speaking for the Environment in Criminal Prosecutions

Sarah Wright

1 Introduction

Eco-justice involves a tripartite conceptualisation of justice regarding protection of humans, the natural environment and non-human species from environmental harms arising from human activities (White, 2014: 15–16). Participation by humans speaking for the environment and non-human species is important to operationalising eco-justice (White & Heckenberg, 2014: 51–53). In the Australian state of New South Wales ('NSW'), the public and environmental groups have a long-standing history of taking civil proceedings to protect the environment, including non-human species (Smith & Higginson, 2022: 253–267). Civil enforcement proceedings in the NSW Land and Environment Court ('NSWLEC') are facilitated through 'open standing' (see e.g. *Environmental Planning and Assessment Act 1979* (NSW) ('*EP&A Act*'): s 9.45; Bates, 2023: 820). In contrast, public involvement in speaking

S. Wright (✉)
University of Wollongong, Wollongong, NSW, Australia
e-mail: sawright@uow.edu.au

R. Killean et al. (eds.), *Conceptualising Justice*, Palgrave Studies in Green Criminology, https://doi.org/10.1007/978-3-032-19791-7_5

for the environment is limited in criminal prosecutions. Some environmental laws only permit government prosecutions (e.g. *Biodiversity Conservation Act 2016* (NSW) ('*BC Act*'): s 13.3). Others restrict private prosecutions to limited circumstances (e.g. *Protection of the Environment Operations Act 1997* (NSW) ('*POEO Act*'): s 219). Private prosecutions, where possible, are rare due to practical difficulties, including lack of investigation powers (Wright, 2024: 575) and high costs. In practice, the public's role in environmental prosecutions has been limited to firstly, providing evidence as a victim or witness *if* asked by a party to give evidence. Secondly, community members have participated in restorative justice ('RJ') conferencing in two NSWLEC prosecutions (*Garrett v Williams* (2007) 151 LGERA 92 ('*Garrett v Williams*'); *Chief Executive, Office of Environment and Heritage v Clarence Valley Council* (2018) 236 LGERA 291 ('*OEH v Clarence Valley Council*')).

In *Environment Protection Authority v Forestry Corporation of NSW* [2024] NSWLEC 84 ('*BECI case*'), the NSWLEC permitted a local environmental group, the Bellingen Environment Centre Inc ('BECI'), to appear as amicus curiae ('a friend of the court') in a sentencing hearing 'essentially… on behalf of endangered koalas' (at [26]). The court has discretion to permit an amicus where they can 'provide assistance… on a matter of fact or law the court would not otherwise obtain' (*BECI case*: [4] citing *Levy v Victoria* (1997) 189 CLR 579 ('*Levy*'): 604). BECI was permitted limited participation to raise certain matters, particularly around impact of the offences (*BECI case*: [17]–[31], [47]). BECI sought to raise concerns regarding the 'social context' of the offences and importance of the environment harmed, including for koala conservation (*BECI case*: [17]–[18]).

While the NSWLEC is a 'problem-solving court' (White, 2017: 122) with wide sentencing powers, permitting an amicus to speak for the environment in a sentencing hearing was novel for the Court. While sentencing was not finalised at the time of writing, the BECI case raises fundamental questions regarding the public's role in speaking for the environment in criminal matters. An amicus may trespass into the prosecutor's discretion in presenting evidence and submissions on behalf of the community and environment. Furthermore, fairness issues arise for defendants due to another advocate making submissions 'against' them (see Williams, 2000: 372, 398 citing Justice Iacobucci, Supreme Court of Canada) and potentially increasing the time and costs involved. Yet there is also the ideal of achieving eco-justice through full voicing of environmental harm (White & Heckenberg, 2014: 51–53). Harm is important in determining the seriousness of offences, appropriate punishment and utilisation of reparative and restorative sentencing outcomes. Full consideration of environmental harm through participatory processes is therefore important to 'doing' eco-justice (White & Heckenberg, 2014: 51–53).

This chapter briefly explains environmental offences, prosecution and sentencing in NSW. It then examines eco-justice in the context of environmental crimes and its operationalisation through participation in prosecutions. Next, it considers the potential role and limitations of the public speaking for the environment by acting as prosecutor, providing evidence as a victim or witness, or participating in RJ conferencing. Lastly, the chapter examines a recent avenue for the public in seeking eco-justice by speaking for the environment in sentencing as amicus, and potential benefits and problems with this approach.

2 Environmental Offences, Prosecution and Sentencing

NSW environmental offences cover areas such as development control, biodiversity, forestry and pollution (see *EP&A Act*; *BC Act*; *Forestry Act 2012* (NSW) ('*Forestry Act*'); *POEO Act*). Prosecution is a high-level sanction (Ayres & Braithwaite, 1992: 35) that, unlike other enforcement responses, may involve participation beyond the regulator and offender, such as by victims and witnesses. While some legislation allows the public to prosecute environmental offences, in practice this task falls to environmental regulators, such as the NSW Environment Protection Authority ('EPA'). More serious offences are prosecuted in the NSWLEC, rather than the Local Court (Wright & Colvin, 2025: 137). The guilty plea rate is high, particularly given that most offences are strict liability offences (see Hamilton, 2019a: 99).

Sentencing considerations derive from sentencing legislation (*Crimes (Sentencing Procedure) Act 1999* (NSW) ('*CSP Act*')), environmental statutes (e.g. *BC Act*: s 13.12) and cases. When sentencing, courts consider the objective seriousness of the offence, offender's subjective circumstances, purposes of sentencing (such as adequate punishment, deterrence and recognising the harm caused) and other relevant principles (for a discussion see Preston, 2011a: 313–345). The sentence must be proportional to the objective seriousness of the offence (*Veen v The Queen (No 2)* (1988) 164 CLR 465: 472, 485–486), including actual or likely environmental harm. Greater harm reflects a more serious offence (see e.g. *POEO Act*: s 241(1)(a); *BC Act*: s 13.12(1)(a); *Environment Protection Authority v Waste Recycling and Processing Corporation* (2006) 148 LGERA 299 ('*EPA v WRAP*'): [145], [148]). Substantial harm constitutes an aggravating factor (*CSP Act*: s 21A(2)(g)).

NSW statutes include broad definitions of the 'environment', covering components of the natural environment (including non-human species), interactions between those components, humans and the built environment (e.g. *POEO Act*: Sch 6). Definitions of 'harm' encapsulate a wide range of behaviours and impacts (e.g. *POEO Act*: Sch 6; *BC Act*: s 1.6(1)). Unsurprisingly, the NSWLEC applies a wide conception of environmental harm in sentencing (*Environment Protection Authority v Orica Australia Pty Ltd* (2014) 206 LGERA 239: [116]). For instance, harm includes impacts on human health and 'quality of life' and 'harm to the environment and its ecology' (*EPA v WRAP:* [145]-[147]). Similarly, 'victims' of environmental crime are broadly identified, including human individuals and groups, the community and the environment and its components, including non-human species (Hamilton, 2021: 26–29; Preston, 2011b: 140–143).

The NSWLEC is a 'problem-solving court', which focuses on human and non-human interests and repairing environmental harm (White, 2017: 122). The Court's problem-solving nature is reflected in the broad range of alternative sentencing orders ('ASOs') it can utilise beyond fines and imprisonment (White, 2018: 261). This includes requiring offenders to 'make good' any damage, pay for an environmental enhancement project, or carry out RJ activities of social or community benefit (see e.g. *BC Act*: pt 13 div 3; *POEO Act*: pt 8.3; Wright, 2018).

3 Seeking Eco-Justice Through Participation in Legal Proceedings

Eco-justice recognises that harm arising from environmental crimes extends beyond humans to ecosystems and other species (White, 2014: 6). It is based upon ideals of justice and protection of 'human, ecological and animal rights' from environmental harms caused by lawful and unlawful human activities (White, 2014: 6, 11–12). White and Heckenberg identify four 'dimensions of justice' within eco-justice:

- *Justice as fairness* concerns distributional justice, namely distribution and sharing of environmental benefits and rights, and distribution of environmental burdens.
- *Justice as acknowledgement* reflects recognition: ensuring respect and 'equal dignity for all' across humans, non-human species and the natural world.
- *Justice as engagement* calls for participation in decision-making processes, including the ability to challenge decisions in court. Given non-human components of the environment, such as animals and trees, cannot speak for themselves, they are given a 'voice' and advocated for by human participants.
- *Justice as functioning* is concerned with the ability of all things to function and flourish (White & Heckenberg, 2014: 50–53 drawing upon Schlosberg, 2007; Preston, 2015: 23–25).

In terms of achieving 'justice' under an eco-justice approach, 'repairing harm and restoration are practical ways in which justice can be done' through the implementation of mechanisms reflecting the four dimensions of justice (White & Heckenberg, 2014: 51–52).

Next, it must be considered how eco-justice is reflected within the NSW criminal justice system, which is concerned with unlawful behaviour. Eco-justice narrows in scope within criminal law because laws permit some environmental harm, such as through approvals and licences (see Hamilton, 2021: 2). Nevertheless, offences seek to uphold legislative objectives aimed at protecting the environment, non-human species and humans from harm (see e.g. the legislative objectives in *POEO Act*:

s 3; *BC Act*: s 1.3). The purposes and principles of sentencing and broad range of sentencing orders aim to: recognise harm to the environment and other victims; punish and hold an offender accountable for the offence and harm; and prevent future crimes (environmental harm) by deterring the offender and others through the penalty imposed (see e.g. *CSP Act*: s 3A; *POEO Act*: s 241, pt 8.3; *BC Act*: pt 13 div 3). The criminal justice system seeks to 'right' the environmental 'wrong' through sentencing orders, such as monetary penalties and environmental restoration or enhancement orders. Accordingly, environmental prosecutions are capable of reflecting White and Heckenberg's conceptions of 'doing justice', particularly through repairing harm and environmental restoration.

(Eco)justice will only be achieved if the sentence is proportional to the seriousness of the offence, including environmental harm. Accordingly, to 'do justice' courts must be fully apprised of environmental harm. Participation by humans speaking for the environment and non-human species regarding any harm and the sentencing outcomes sought to be achieved therefore becomes important to operationalising eco-justice in prosecutions. The ability to take civil proceedings has allowed the public to provide a 'voice' for the environment. For example, a community group concerned about a critically endangered bird successfully argued a development consent was invalid because of failure to submit a species impact statement (*Friends of Tumblebee Inc v ATB Morton Pty Ltd (No 2)* [2016] NSWLEC 16). As discussed below, in NSW, it is generally a government prosecutor that gives a 'voice' to the environment in criminal proceedings through its case (Hamilton, 2021: 31); the public's ability to speak for the environment has been limited.

4 The Public's Role in Speaking for the Environment in Prosecutions

This part examines the public's role in protecting and speaking for the environment in prosecutions by acting as prosecutor, providing evidence as a 'victim' or witness of environmental crime and through participation in RJ conferencing. This reflects the potential roles for the public in prosecutions before the BECI case, which provided another avenue to speak for the environment. Examining these roles and their limitations is necessary to understand why BECI sought to appear as amicus in a sentencing hearing. The role of amici and the BECI case is examined in the subsequent part.

A. *The Public as Prosecutor*

The most expansive role the public or victims can play in criminal proceedings is as the prosecutor, who determines the charges, evidence and submissions placed before the court. In NSW, the prosecutor decides which ASOs to seek, such as diversion of monetary penalties to environmental projects (e.g. creek restoration works) or to an organisation (e.g. a koala animal hospital) (*POEO Act*: s 250(1)(e); *Environment Protection Authority v Elf Farm Supplies Pty Ltd* [2017] NSWLEC 60: [108]; *Environment Protection Authority v Koppers Carbon Materials & Chemicals Pty Ltd* [2021] NSWLEC 13: [126]). While the court can impose several ASOs of its own volition, it relies on the parties to propose the form of some ASOs, including environmental projects (Hamilton, 2021: 45; Wright, 2018: 616–618). Accordingly, overall, it is the prosecutor who speaks for the environment in prosecutions and plays a key role in ensuring environmental protection, remediation and enhancement through sentencing orders and consequently, achieving eco-justice.

Environmental crimes are prosecuted virtually exclusively by government environmental regulators. The potential for private prosecutions by the public is restricted by legislation and/or practical difficulties. The starting point regarding who can prosecute is the 'common informer' provision which permits any person to prosecute unless that right is restricted by legislation creating the offence (*Criminal Procedure Act 1986* (NSW): s 14). The power to prosecute NSW environmental crimes generally falls into three categories:

(1) Legislation creating the offence does not restrict who can prosecute (e.g. planning law offences: *EP&A Act*).

(2) Specified government bodies or officials can prosecute. Additionally, any person may prosecute with the court's leave if certain criteria are met (e.g. pollution offences: *POEO Act*: ss 217–220).
(3) Only specified government bodies or officials can prosecute (e.g. biodiversity and forestry operations offences: *BC Act*: s 13.3).

Private environmental prosecutions are exceptionally rare in NSW, with one known NSWLEC case (*M.&S. Investments (NSW) Pty Ltd v Affordable Demolitions and Excavations (No 3)* [2025] NSWLEC 30: [121]). Where leave is required for private prosecutions, provisions prevent the court granting leave if the regulator has taken some other 'relevant action', such as issuing a clean-up notice or exercise of other regulatory powers (*POEO Act*: s 219; Wright, 2024: 574). This limits the scope for private prosecutions to be taken.

Private prosecutions also face practical problems including private parties lacking investigation powers afforded to government regulators, the high evidentiary burden (Wright, 2024: 575) and significant cost of legal representation and expert evidence. Accordingly, criminal prosecution essentially remains 'the sole domain of government regulators' (Bates, 2023: 744).

B. *The Public as Victims and Witnesses*

The role the public plays as a witness in environmental prosecutions, including as a victim or representative of the environment as victim, is largely determined by government prosecutors. The public may have: witnessed the crime; noticed resulting environmental harm, such as dead fish in a river; or been impacted as a 'victim', such as by offensive odours. The prosecutor may rely on such evidence to establish the crime or harm caused. However, beyond witness statements regarding what community members or victims observed or experienced, courts typically rely on expert evidence given by persons such as scientists to establish the extent of environmental harm because often technical issues are involved, such as how long the harmed environment may take to regenerate.

While witness statements from community members may be needed to establish an environmental prosecutor's case, in practice, the role of witnesses may be limited. There is a high guilty plea rate in environmental offences (Hamilton, 2019a: 99). Sentencing often proceeds on a statement of facts agreed between the prosecutor and defendant (Land and Environment Court of NSW, 2020: [36]). The prosecutor may not lead other evidence or only limited evidence on disputed matters, such as expert evidence on environmental harm. Any 'voice' given to human or environmental victims may therefore be presented through wording agreed by the parties or expert evidence. That 'voice' may be lost and victims' experiences 'sanitised' by the objective way facts and evidence are presented (see Hamilton, 2021: 32–35).

NSW legislation permits a victim impact statement ('VIS') to be tendered by the prosecutor, read by a victim (or representative) in a sentencing hearing and considered by the court (*CSP Act*: pt 3, div 2). However, those provisions do not apply to NSWLEC environmental prosecutions (*CSP Act*: s 27). They apply, for instance, to certain general criminal offences resulting in a person's death or actual physical bodily harm (*CSP Act*: s 27). Accordingly, no legislative provision expressly permits a VIS to be read on behalf of a person or the environment in environmental sentencing, despite legislative recognition that environmental offences may cause death or serious injury to a person or significant environmental harm (e.g. *EP&A Act*: s 9.52(2)(b); *Local Land Services Act 2013* (NSW): s 60N(1)).

In 2003, the NSW Law Reform Commission ('NSWLRC') raised the issue of VISs in environmental sentencing (NSWLRC, 2003: 194). The NSWLRC noted the court must consider any environmental harm (NSWLRC, 2003: 195). Furthermore, '[s]uch matters could conceivably be brought to the attention of the court by means of a VIS or other victim report prepared by either the victim or a representative of a group of victims' (NSWLRC, 2003: 195). From the context of the Commission's report, these comments referred to human victims, not the environment or non-human species (NSWLRC, 2003: 193–195). Formal VISs have not been adopted within NSWLEC sentencing.

Jurisdictions including Hong Kong, South Africa and Zambia introduced species VISs in areas such as wildlife crime to give a 'voice' to animals, recognise them as victims in their own right and the need to protect their welfare and intrinsic value (MacBeath & Whitford, 2024:

3, 5, 7–11). Species VISs were adopted to assist courts to properly assess environmental harm due to concerns regarding 'overly lenient sentences' (MacBeath & Whitford, 2024: 4–7). Species VISs in illegal wildlife trade prosecutions have, for example, taken the form of expert testimony or standard scientifically-based expert statements for individual species contained in a database that can be utilised when needed (MacBeath & Whitford, 2024: 7–12).

As discussed, the NSWLEC applies a broad meaning of environmental harm in sentencing that considers ecocentric and anthropocentric concerns. The prosecution leads expert evidence to establish environmental harm, including to non-human species. While NSW does not employ species VIS, expert evidence currently fulfils this role.

C. *Restorative Justice Conferencing*

A NSWLEC practice note recognises the availability of RJ conferencing for environmental offences and that the defendant can agree to RJ activity orders when sentenced by the court (Land and Environment Court of NSW 2020: [28]). RJ conferencing involves 'the defendant and victims (people and the environment)' as willing participants (Land and Environment Court of NSW, 2020: [28]) who 'come together to resolve collectively how to deal with the aftermath of the offence and its implications for the future' (Hamilton, 2021: 78 citing Marshall, 1996: 37). RJ conferencing represents one way of operationalising eco-justice: it provides a participatory process allowing the voices of victims and other stakeholders to be heard and for the participants to agree on restorative outcomes to repair any harm (Hamilton, 2021: 78–80; White & Heckenberg, 2014: 52), such as through environmental or community projects and apologies. However, RJ conferencing in NSW environmental prosecutions is rare, with use in only two cases involving Aboriginal cultural heritage offences (*Garrett v Williams*; *OEH v Clarence Valley Council*; for a discussion see Hamilton, 2008, 2019b). There is no requirement that the parties consider using RJ conferencing (Hamilton, 2019b: 204–205). While the court may suggest RJ conferencing be considered (see *Garrett v Williams*: [40]), participation is voluntary (Hamilton, 2021: 80). Its use is therefore dependent on awareness of its benefits and willingness of defendants and victims to participate (Hamilton, 2019b: 204–205). Uptake has been low.

D. ***Conclusion on Existing Opportunities to Participate***

In summary, currently few opportunities exist for the public to participate in environmental sentencing and therefore, seeking eco-justice. Private prosecutions are virtually non-existent. Victims, including representatives for the environment as victim, are unlikely to participate during sentencing. Firstly, because there is no provision for VISs. Secondly, environmental harm is often presented through an agreed statement of facts and expert evidence. The 'voice' of victims may be lost or sterilised through this process and, as Hamilton (2021: 32) notes, '[t]he effectiveness of the voice given to… victims is contingent on the prosecution actually identifying them'. While eco-justice can be achieved through RJ conferencing, its use is rare.

5 An Additional Role for the Public in Environmental Prosecutions: Appearing as Amicus Curiae

This part examines a recent approach utilised by a community group: appearing as amicus to speak for the environment in a sentencing hearing. It firstly considers when amici may be permitted. Next, the BECI case is examined to consider matters the Court permitted BECI to address regarding sentencing. Lastly, this part reflects on the case in terms of achieving eco-justice and concepts within the criminal justice system.

A. ***Amicus Curiae and Criminal Proceedings***

In Australia, permission to be heard as amicus is at the court's discretion (*Levy*: 604). The accepted basis to permit an amicus is that the 'person is willing to offer the Court a submission on law or relevant fact which will assist the Court in a way [it] would not otherwise have been assisted'

(*Levy*: 604, citations omitted). The courts consider 'all the circumstances' (*National Australia Bank Ltd v Hokit Pty Ltd* (1996) 39 NSWLR 377: 381 ('*Hokit*')) and take a cautious approach to maintain 'the efficient operation of the Court' (*Kruger v Commonwealth* [1996] HCATrans 68 ('*Kruger*')). The court should consider 'whether the appearance is in the interests of the parties', ensure the court's time is not occupied unnecessarily (*Carson v Legal Services Commissioner* [2000] NSWCA 308: [289]; *Hokit*: 381) and that any delay or increased costs for the parties is 'not disproportionate to the assistance that is expected' (*Levy*: 605).

The court defines the extent to which an amicus can participate (*BECI case*: [4]), including whether they can make only written or also oral submissions (see Kenny, 1998: 170). An amicus is not a party and has 'no entitlement to lead evidence' (*Bropho v Tickner* (1993) 40 FCR 165: 172 ('*Bropho*')). However, 'it may sometimes be appropriate to allow an amicus curiae to complete the evidentiary mosaic', but not such as to allow it to 'hijack the parties' case' (*Bropho*: 172–3). In prosecutions, amicus evidence is usually only allowed 'in exceptional circumstances' (*BECI case*: [45]).

Amicus applications are rarer in Australia than other countries (*Bropho*: 172) and in criminal versus civil matters, with few NSWLEC criminal cases (*BECI case*: [6]). This author located only two other NSWLEC prosecutions involving amici (search conducted on the term 'amicus' within NSWLEC judgments via AustLII: https://www.austlii.edu.au/ on 14 June 2025). One case involved a barrister assisting an unrepresented defendant (*Garrett v Langmead* [2006] NSWLEC 627: [43]). In the other case, a solicitor was permitted to be heard as amicus regarding whether service upon them effected service upon the defendant (*Filipowski v Frey* [2004] NSWLEC 182: [3]-[4]).

In the BECI case, the NSWLEC noted (without discussing) four non-environmental criminal matters the parties referred to (at [8]). In those cases, a government body sought to appear as amicus. The three cases where this was permitted involved issues regarding whether criminal prosecutions could be maintained (*Director of Public Prosecutions v Yigit* [2008] NSWCA 226; *R v Seller* [2012] NSWSC 934) and a constitutional challenge to mandatory minimum sentencing provisions (*Karim v*

The Queen (2013) 83 NSWLR 268). In the fourth case, where a government body sought to appear as amicus in a sentencing hearing to raise international human rights law arguments, the Court refused the application noting well-established sentencing principles applied (*R v GJ* [2005] NTCCA 20: [47], [65]). Other NSW criminal cases that permitted amici included cases with self-represented defendants where an amicus assisted the court (*AD v Commissioner of the Australian Federal Police* [2018] NSWCA 89: [2]) or defendant (e.g. *R v Wiggins (No 2)* [2017] NSWSC 348: [13]), or acted as 'contradictor to the Crown' regarding mental illness and defences (*R v Waszczuk* [2012] NSWSC 1080: [5]). Permitting an amicus ensured fairness to the defendant (Mirsane, 2022: 693–694). The criminal amici cases were different to the BECI case where an environmental group was permitted to make submissions and lead evidence at a sentencing hearing on behalf of the environment.

B. *The BECI Case*

The Forestry Corporation of NSW ('FC') is a statutory State-owned corporation (*Forestry Act*: s 5). It pleaded guilty to four offences prosecuted in the NSWLEC under the *Forestry Act* involving breach of approval conditions, namely failure to retain specified trees when carrying out forestry operations within a State forest (*BECI case*: [2]). The right to prosecute was restricted to government prosecutors, consequently the EPA prosecuted as regulator of FC's forestry operations (*Forestry Act*: s 69SB; *BC Act*: s 13.3). The EPA had not proposed to call BECI members to give evidence, nor was RJ conferencing planned.

Prior to the sentencing hearing, BECI requested to appear as amicus, provide evidence and make submissions on sentencing (*BECI case*: [3], [5]). BECI is:

> a voluntary non-government organisation which brings together people in the Bellingen shire interested in protecting the environment. Its aims include the promotion of environmental protection and biodiversity in the Bellingen Valley, and it conducts a number of activities to achieve its objectives. (*BECI case*: [5])

BECI had lobbied the NSW government to create a Great Koala National Park ('GKNP') to protect koalas, which was proposed to include the State forest where the offences occurred (*BECI case*: [18]). Although the government made an election promise to create a GKNP, it had not yet been established, nor its area finally determined (*BECI case*: [18], [23]). Koalas are 'facing a very high risk of extinction… in the near future' and are listed as endangered (*BECI case*: [24]; *BC Act*: s 4.4(2)(a), sch 1; *Environment Protection and Biodiversity Conservation Act 1999* (Cth): ss 178, 179(4)(b)). Koala habitat has been impacted by clearing vegetation for human activities, bushfires and climate change (Department of Agriculture, Water and the Environment, 2022: 5).

The parties opposed BECI's application, arguing it would 'not assist the Court and will lead to disproportionate costs' (*BECI case*: [3], [6]). The defendant raised: that the right to prosecute was limited to government prosecutors; there was no legislative right for others to intervene, participate or be heard; the criminal procedure legislation did not contemplate involvement of parties beyond the prosecutor and defendant, including regarding conducting the case and the right to call witnesses; the prosecutor must establish aggravating factors, including substantial environmental harm, beyond reasonable doubt and relatedly, the VIS provisions do not apply to environmental offences (*BECI case*: [11]–[14]). The EPA argued it had sole discretion as prosecutor to decide which witnesses to call in establishing aggravating factors including environmental harm (*BECI case*: [16]).

BECI sought to participate regarding multiple sentencing considerations. The central issue related to the impact of the offences on the proposed GKNP:

> The BECI seeks to assist the Court by providing the local social context within which the offences occurred given the significance of the area damaged by the offences to the local community and the community's views of the seriousness of the offences. (*BECI case*: [17])

BECI's concerns related to the impacts on long-term koala survival and conservation values of the proposed GKNP, for which it had been an avid proponent (*BECI case*: [18]). BECI sought to lead evidence and make submissions on 'the ecological and social significance of the damaged area', arguing it had 'special knowledge' due to its involvement in the proposed GKNP (*BECI case*: [19]). The matters BECI wished to raise were said to be relevant to several sentencing factors, including environmental harm (*BECI case*: [19]–[20]). The EPA argued it would lead evidence and submissions on relevant sentencing matters, leaving nothing for BECI to assist the Court with, as required by the amicus test (*BECI case*: [21]). The parties raised concerns regarding reports BECI sought to rely on, relating to the status of koala populations and the GKNP proposal and its underlying scientific rationale (*BECI case*: [19]–[23]). They argued that as prior existing reports, rather than reports addressing environmental harm from the offences, their relevance was unclear (*BECI case*: [23]). Nor were they prepared 'in accordance with the expert witness code of conduct' (*BECI case*: [23]). Furthermore, the GKNP was only at the proposal stage, which the EPA argued 'diminishes the utility of the BECI's potential advice as amicus' (*BECI case*: [23]).

In determining whether to permit BECI to make submissions on the impacts of the offences, Pain J noted 'environmental protection is a matter of interest and concern to the whole community' (*BECI case*: [24]). Her Honour noted BECI's longstanding environmental protection role and involvement with the proposed GKNP stemming from concerns 'to provide much needed habitat for koalas', an endangered and 'iconic Australian species' (*BECI case*: [24]). While Pain J acknowledged the EPA is 'an experienced regulator', that would 'no doubt undertake its duty responsibly' to raise relevant sentencing factors, her Honour also noted that it was unknown how the prosecutor would raise community concerns or if it would address the proposed GKNP (*BECI case*: [25], [30]). Pain J opined:

> Community interest and concern in the preservation of the environment is potentially wider than an environmental regulator may conceive it... [T]he BECI essentially wishes to appear on behalf of endangered koalas in the Bellingen region in having matters placed before the Court which could be relevant to sentencing.
>
> A wider context for considering this application is increasing recognition of the importance of restorative justice in environmental crime in many

> jurisdictions... Essentially restorative justice requires that all aspects of the environment including human and non-human relationships with it need to be recognised in order for a court to consider all aspects of environmental harm when sentencing and to craft appropriate orders in response. (*BECI case*: [26]–[27])

Her Honour noted the availability of RJ conferencing between a defendant, prosecutor and impacted community members and that defendants can agree to carry out RJ activity orders of social or community benefit (*BECI case*: [28]–[29]). Her Honour identified RJ conferencing and activity orders as a way 'BECI could have been engaged constructively' in the sentencing process (*BECI case*: [29]). Pain J concluded:

> BECI has a unique perspective concerning the environmental values of the region over which the GKNP is proposed which may assist the Court and should be before the sentencing judge. In this case it appears warranted to inform the Court about the GKNP proposal of which the BECI is the proponent and place the area where the offences occurred in that context. (*BECI case*: [30])

BECI was allowed to file three reports regarding environmental harm, with the trial judge to determine whether they could be relied on (*BECI case*: [31], [47]; *Environment Protection Authority v Forestry Corporation of New South Wales (No 2)* [2025] NSWLEC 24: [1] ('*BECI evidence case*')). In a later hearing, the EPA and defendant again raised lack of relevance to sentencing given the reports did not address environmental harm from the offences. Pain J prohibited BECI from relying on one report and only a few pages of the other reports were permitted (*BECI evidence case*: [29]–[31]).

Some further matters BECI sought to make submissions on were not permitted, including appropriate ASOs. Pain J held the EPA could propose appropriate ASOs (*BECI case*: [43]). BECI also sought to 'provide evidence and submissions of local community outrage' regarding the offences given FC's prior record (*BECI case*: [32]). The Court concluded the EPA and BECI should discuss how the EPA would approach this matter before the Court determined this point (*BECI case*: [33]).

C. *Reflections on the BECI Case*

The BECI case is consistent with notions underlying eco-justice, including ensuring 'justice is done' through full representation of concerns regarding environmental harm, which helps ensure sentences are proportional to the seriousness of an offence. The case operationalised eco-justice by allowing BECI's participation to voice those concerns by speaking for koalas and the environment. However, once BECI was permitted to make submissions on environmental harm, arguably greater consideration of whether BECI could make submissions on appropriate ASOs was required. Refusing this request raises inconsistencies with reasoning permitting BECI to make submissions on environmental harm. Pain J recognised BECI's longstanding history regarding local environmental protection, its 'unique perspective' on local environmental values and the importance of 'consider[ing] all aspects of environmental harm when sentencing and to craft appropriate orders in response' (*BECI case*: [24]–[27], [30]). Repairing harm and environmental restoration, such as through environmental projects, are a means of achieving eco-justice (White & Heckenberg, 2014: 51–52). In NSW, monetary penalties can be diverted to beneficial projects that enhance the environment through an ASO (see e.g. *BC Act*: ss 13.19; 13.25(1); Wright, 2018: 616). These projects do not remediate harm caused by an offence (for which another ASO is available); they instead fund additional initiatives providing reparation for the offence (Wright, 2018: 616, 623). The Court stated the EPA could adequately make submissions on ASOs (*BECI case*: [43]), but did not indicate if the EPA had determined which orders to seek. The EPA does not always seek diversion of monetary penalties to environmental projects, resulting in a fine being imposed, as occurred in some previous FC prosecutions (e.g. *Environment Protection Authority v Forestry Corporation of New South Wales* [2024] NSWLEC 78). Where an environmental project order is sought, the parties may disagree on the appropriate project (e.g. *Environment Protection Authority v Austar Coal Mine Pty Ltd* [2011] NSWLEC 252: [48]–[61]). Some projects have a stronger nexus to the harm caused and the local area (Hamilton, 2021: 45–49; Wright, 2018: 623). BECI, a community group involved in local environmental protection, arguably may have provided valuable assistance on these matters. Participation by speaking for the environment in 'crafting appropriate orders' is a means of operationalising eco-justice and achieving sentencing outcomes which better reflect restorative and reparative justice.

While there are arguably benefits in permitting amici to speak for the environment at sentencing, it also raises potential concerns. Firstly, it 'might trigger a floodgate of "busybodies"' (Durbach, 1998: 178) in other prosecutions. Fears around 'opening the floodgates' in other environmental contexts, such as open or extended standing provisions in civil proceedings, have been dismissed as unrealised (Bates, 2023: 820; Macintosh et al., 2017: 109; Pain & Wright, 2003: 6).

The Bylong Valley Protection Alliance ('BVPA') community group sought joinder as a party in NSWLEC judicial review proceedings (*KEPCO Bylong Australia Pty Ltd v Independent Planning Commission* [2020] NSWLEC 38 ('*KEPCO*')). The applicant company challenged the Independent Planning Commission's decision to refuse development consent to its proposed mine. BVPA played a substantial role as lead objector to the mine in the Commission proceedings, providing evidence and written and oral submissions (*KEPCO*: [14], [23], [80]). While there were many other objector participants, none requested involvement in the NSWLEC proceedings, nor did they meet the joinder test (*KEPCO*: [79]–[80]). In permitting BVPA's joinder, Moore J opined that '[a]lthough the floodgates might be opened, there is no evidence [of] even a minute trickle flowing through them as a result' (*KEPCO*: [79]).

Similar points dismissing floodgates arguments could be made regarding permitting an amicus to speak for the environment in a sentencing hearing. The BECI case is a pertinent example. BECI had a longstanding history of local forest protection with heavy involvement as 'one of the primary originators' of the proposed GKNP which seeks to protect koalas and had a 'unique perspective' on local environmental values (*BECI case*: [18], [30]). No one else sought to appear as amicus. Arguably, like BVPA in the *KEPCO case*, BECI may have been in a unique position.

Secondly, permitting an environmental group to intervene in sentencing raises broader concerns about the operation of the criminal justice system. It involves potential trespass into the prosecutor's discretion in presenting the evidence and submissions on behalf of the community and environment. There are also issues of fairness to the parties, particularly defendants. Fairness is relevant to justice (Hamilton, 2021: 58). A traditional criminal case in common law systems involves two parties: prosecutor and defendant. A 'prosecutor's principal role is to assist the court to arrive at the truth and to do justice between the community and the accused' (NSW Office of the Director of Public Prosecutions, 2021: 13). Prosecutors have ethical obligations (e.g. *Legal Profession Uniform Law Australian Solicitors' Conduct Rules 2015* (NSW) r 29). They must act fairly, impartially and with detachment (Gibson, 2017: 264–265; NSW Office of the Director of Public Prosecutions, 2021: 13). They have a responsibility to raise aggravating factors (Gibson, 2017: 287), including substantial environmental harm. These are matters adverse to the accused which increase the penalty. The role of amici is also to assist the court (*Levy*: 604). It has been accepted 'that [an] amicus may enter into partisan advocacy' (Kenny, 1998: 161), although the precise role and any ethical obligations an amicus voicing concerns on behalf of the environment in a sentencing hearing owes to a defendant are ill-defined. With the addition of an amicus, a defendant may find the 'balance' is tipped, with more than one advocate making submissions 'against' them (see Williams, 2000: 398 citing Justice Iacobucci, Supreme Court of Canada) on matters which may increase a sentence, such as environmental harm. While the defendant in the BECI case was a legally represented government emanation, this will not always be the case. While general principles regarding permitting amici raise the need for a cautious approach (*Levy*: 604 citing *Kruger*), arguably greater caution is needed where an amicus seeks to make submissions against a defendant in sentencing.

Where 'uninvited guests' are permitted, the parties face potential broadening of the issues and additional time and costs (Kenny, 1998: 167). Usually amici do not pay costs (*Muswellbrook Shire Council v Hunter Valley Energy* [2019] NSWCA 216: [67]). Defendants usually pay prosecutors' costs in NSW environmental prosecutions. Concerns regarding 'unnecessary costs' may be heightened where amici appear without legal representation, as occurred with BECI (*BECI case*: [46] citing defendant's submissions). However, the Court noted 'BECI's [limited] participation will be closely defined' and supervised to prevent 'disproportionate costs' being incurred (*BECI case*: [46]–[47]). Nevertheless, issues relating to the amicus resulted in two additional hearings prior to sentencing. The sentencing hearing occurred nine months after originally scheduled (see *BECI case*: [1]; *BECI evidence case*: [1]). The pertinent question is when additional costs, time and delay are considered 'disproportionate' compared to seeking to better voice environmental concerns. Allowing BECI to submit a VIS may have proven more efficient and cost-effective. However, any proposal to introduce VISs would need careful consideration regarding: whether VISs are necessary; what VISs may address given prosecutors usually lead expert evidence on environmental harm; eligibility to make a VIS on behalf of the environment; the efficient operation of the courts; and costs implications.

6 Conclusion

Participation by the public to speak for the environment in sentencing hearings has been limited. In the BECI case, the NSWLEC provided a new avenue for public participation to voice those concerns, thereby operationalising eco-justice in practice. As Kirby P noted in a different context relating to amici: '[v]iews differ upon procedural innovation. But the courts should not forfeit the adaptation of their procedures to slavish adherence to methodologies adopted … in different times for different problems' (*Breen v Williams* (1994) 35 NSWLR 522: 533). The NSWLEC is a 'problem-solving' court (White, 2017: 122) that has embraced many innovations in its role in upholding environmental protection legislation, including sentencing offenders. Permitting an amicus to speak on behalf of koalas in a sentencing hearing is one of its latest innovations. How this will impact the sentencing decision and whether a flood of similar applications will follow remains to be seen. Nevertheless, the BECI case provokes further consideration regarding whether and how victims, including those seeking to speak for the environment, can and should be allowed to participate in sentencing. Further consideration of potential vehicles to do so, including as amicus, requires fuller examination to ensure that the balancing of environmental concerns is properly weighed against other fundamental aspects of the criminal justice system. This includes fairness to defendants, efficient use of the court's time given expert evidence on environmental harm is already utilised and cost implications for the parties. Other avenues, such as VISs, may ultimately provide a better option for all stakeholders, but these issues require further examination.

References

AD v Commissioner of the Australian Federal Police [2018] NSWCA 89.

Ayres, I., & Braithwaite, J. (1992). *Responsive Regulation: Transcending the Deregulation Debate.* Oxford University Press.

Bates, G. (2023). *Environmental Law in Australia* (11th ed.). LexisNexis.

Biodiversity Conservation Act 2016 (NSW).

Breen v Williams (1994) 35 NSWLR 522.

Bropho v Tickner (1993) 40 FCR 165.

Carson v Legal Services Commissioner [2000] NSWCA 308.

Chief Executive, Office of Environment and Heritage v Clarence Valley Council (2018) 236 LGERA 291.

Crimes (Sentencing Procedure) Act 1999 (NSW).

Criminal Procedure Act 1986 (NSW).

Department of Agriculture, Water and the Environment. (2022). *Conservation Advice for Phascolarctos cinereus (Koala) Combined Populations of Queensland, New South Wales and the Australian Capital Territory.*

Director of Public Prosecutions v Yigit [2008] NSWCA 226.

Durbach, A. (1998). Interveners in High Court Litigation: A Comment. *Adelaide Law Review, 20*, 177–182.

Environment Protection Authority v Austar Coal Mine Pty Ltd [2011] NSWLEC 252.

Environment Protection Authority v Elf Farm Supplies Pty Ltd [2017] NSWLEC 60.

Environment Protection Authority v Forestry Corporation of New South Wales [2024] NSWLEC 78.

Environment Protection Authority v Forestry Corporation of New South Wales (No 2) [2025] NSWLEC 24.

Environment Protection Authority v Forestry Corporation of NSW [2024] NSWLEC 84.
Environment Protection Authority v Koppers Carbon Materials & Chemicals Pty Ltd [2021] NSWLEC 13.
Environment Protection Authority v Orica Australia Pty Ltd (2014) 206 LGERA 239.
Environment Protection Authority v Waste Recycling and Processing Corporation (2006) 148 LGERA 299.
Environment Protection and Biodiversity Conservation Act 1999 (Cth).
Environmental Planning and Assessment Act 1979 (NSW).
Filipowski v Frey [2004] NSWLEC 182.
Forestry Act 2012 (NSW).
Friends of Tumblebee Inc v ATB Morton Pty Ltd (No 2) [2016] NSWLEC 16.
Garrett v Langmead [2006] NSWLEC 627.
Garrett v Williams (2007) 151 LGERA 92.
Gibson, R. (2017). *Prosecuting.* Thomson Reuters.
Hamilton, M. (2008). Restorative Justice Intervention in an Environmental Law Context: Garrett v Williams, Prosecutions under the Resource Management Act 1991 (NZ), and Beyond. *Environmental and Planning Law Journal, 25*(4), 263–271.
Hamilton, M. (2019a). *Restorative Justice Conferencing in Response to Pollution Offending: A Vehicle for the Achievement of Justice as Meaningful Involvement* (PhD Dissertation). UNSW.
Hamilton, M. (2019b). Restorative Justice Intervention in an Aboriginal Cultural Heritage Protection Context: Chief Executive, Office of Environment and Heritage v Clarence Valley Council. *Environmental and Planning Law Journal, 36*(3), 197–211. https://doi.org/20.500.14802/6889
Hamilton, M. (2021). *Environmental Crime and Restorative Justice: Justice as Meaningful Involvement.* Palgrave Macmillan. https://doi.org/10.1007/978-3-030-69052-6
Karim v The Queen (2013) 83 NSWLR 268.
Kenny, S. (1998). Interveners and Amici Curiae in the High Court. *Adelaide Law Review, 20*, 159–171.
KEPCO Bylong Australia Pty Ltd v Independent Planning Commission [2020] NSWLEC 38.
Kruger v Commonwealth [1996] HCATrans 68.
Land and Environment Court of NSW (2020). *Practice Note - Class 5 Proceedings.*
Legal Profession Uniform Law Australian Solicitors' Conduct Rules 2015 (NSW).

Levy v Victoria (1997) 189 CLR 579.
Local Land Services Act 2013 (NSW).
M.&S. Investments (NSW) Pty Ltd v Affordable Demolitions and Excavations (No 3) [2025] NSWLEC 30.
MacBeath, A., & Whitford, A. (2024). *Species Victim Impact Statements: Giving a Voice to the Unheard Victims of Environmental Crime.* Global Initiative Against Transnational Organized Crime.
Macintosh, A., Roberts, H., & Constable, A. (2017). An Empirical Evaluation of Environmental Citizen Suits under the Environment Protection and Biodiversity Conservation Act 1999 (Cth). *Sydney Law Review, 39*, 85–121.
Marshall, T. F. (1996). The Evolution of Restorative Justice in Britain. *European Journal on Criminal Policy and Research, 4*(4), 21–43. https://doi.org/10.1007/BF02736712
Mirsane, M. (2022). The Role of Amicus Curiae (Friend of the Court) in Judicial Systems with Emphasis on Canada and Alberta. *Alberta Law Review, 59*(3), 669–699. https://doi.org/10.29173/alr2690
Muswellbrook Shire Council v Hunter Valley Energy [2019] NSWCA 216.
National Australia Bank Ltd v Hokit Pty Ltd (1996) 39 NSWLR 377.
NSW Law Reform Commission. (2003). *Sentencing: Corporate Offenders* (Report No 102).
NSW Office of the Director of Public Prosecutions (2021). *Prosecution Guidelines.*
Pain, N., & Wright, S. (2003, October 24). The Rise of Environmental Law in New South Wales and Federally: Perspectives from the Past and Issues for the Future. In *National Environmental Law Association Annual Conference.* Broken Hill.
Preston, B. J. (2011a). Principled Sentencing for Environmental Offences. In L. Paddock, D. Qun, L. J. Kotzé, D. L. Markell, K. J. Markowitz & D. Zaelke (Eds.), *Compliance and Enforcement in Environmental Law: Towards More Effective Implementation* (pp. 313–366). Edward Elgar.
Preston, B. J. (2011b). The Use of Restorative Justice for Environmental Crime. *Criminal Law Journal, 35*, 136–153.
Preston, B. J. (2015). The Effectiveness of the Law in Providing Access to Environmental Justice: An Introduction. In P. Martin, S. Z. Bigdeli, T. Daya-Winterbottom, W. du Plessis & A. Kennedy (Eds.), *The Search for Environmental Justice* (pp. 23–43). Edward Elgar.
Protection of the Environment Operations Act 1997 (NSW).
R v GJ [2005] NTCCA 20.
R v Seller [2012] NSWSC 934.

R v Waszczuk [2012] NSWSC 1080.
R v Wiggins (No 2) [2017] NSWSC 348.
Schlosberg, D. (2007). *Defining Environmental Justice: Theories, Movements, and Nature*. Oxford University Press.
Smith, J., & Higginson, S. (2022). 'Please Come In': Access to Justice and the Development of Jurisprudence in the Land and Environment Court of NSW. In E. Fisher & B. Preston (Eds.), *An Environmental Court in Action: Function, Doctrine and Process* (pp. 253–267). Hart.
Veen v The Queen (No 2) (1988) 164 CLR 465.
White, R. (2014). *Environmental Harm: An Eco-Justice Perspective.* Policy Press.
White, R. (2017). Reparative Justice, Environmental Crime and Penalties for the Powerful. *Crime, Law and Social Change, 67*(2), 117–132. https://doi.org/10.1007/s10611-016-9635-5
White, R. (2018). Sentencing Environmental Offenders: It Is Not Just About the Money. In T. Spapens, R. White, D. van Uhm & W. Huisman (Eds.), *Green Crimes and Dirty Money* (pp. 250–265). Routledge.
White, R., & Heckenberg, D. (2014). *Green Criminology: An Introduction to the Study of Environmental Harm.* Routledge.
Williams, G. (2000). The Amicus Curiae and Intervener in the High Court of Australia: A Comparative Analysis. *Federal Law Review, 28*(3), 365–402. https://doi.org/10.22145/flr.28.3.2
Wright, S. (2018). Re-Examining the Approach to Alternative Sentencing Orders in New South Wales Pollution Law. *Environmental and Planning Law Journal, 35*, 606–634.
Wright, S. (2024). Pollution Control and Waste Disposal. In P. Williams (Ed.), *The Environmental Law Handbook: Legislation, Case Law and Policy in NSW* (pp. 483–608). Thomson Reuters.
Wright, S., & Colvin, V. (2025). Enforceable Undertakings as an Alternative to Prosecution in Australian Environmental Law Enforcement. In R. Pereira & T. Fajardo (Eds.), *A Research Agenda for Environmental Crime and the Law* (pp. 133–161). Edward Elgar.

6

Justice as Meaningful Involvement: Differentiating Legal Versus Illegal Harm and the Role of Restorative Justice

Mark Hamilton

1 Introduction

Justice as it concerns environmental issues takes many forms, as this book attests. Pertinent examples include environmental justice which is concerned with humans, particularly with how the benefits and burdens of environmental exploitation are shared (Murdock, 2021); climate justice which is concerned with the 'maldistribution of climate change impacts…' on individuals and societies (Porter et al., 2020: 293); ecological justice which is concerned with decision making on ecosystems (for an overview see White, 2008); species and multispecies justice which are concerned with the effects of environmental harms on animal and plant species (for an overview see Celermajer et al., 2021; Goyes & Sollund, 2018), including concerns with invertebrate species, which is encapsulated in invertebrate justice (see Durrant, Chapter 9 in this volume).

M. Hamilton (✉)
Australian Catholic University, Sydney, NSW, Australia
e-mail: mark.hamilton@acu.edu.au

R. Killean et al. (eds.), *Conceptualising Justice*, Palgrave Studies in Green Criminology, https://doi.org/10.1007/978-3-032-19791-7_6

Other conceptualisations of justice draw on multiple frames. For example, eco-justice draws on environmental justice (the victim is human), ecological justice (the victim is specific environments) and species justice (the victim is animals and plants) to frame a more expansive conceptualisation of justice (see White, 2018 and Chapter 2 in this volume). Another example of an expansive conceptualisation of justice is justice as meaningful involvement (Hamilton, 2019b, 2021). Such a conceptualisation entails three components—recognition, participation and capability, and when used following environmental offending, restorative justice conferencing is claimed as a vehicle for its achievement. Notwithstanding this conceptualisation in the context of environmental offending, green criminology acknowledges the harm to the environment which can also be caused by legal activity. Therefore, an assessment is needed as to what justice as meaningful involvement looks like in the face of legal activity which harms the environment, and what vehicle(s) can achieve justice so conceived. This is the original contribution this chapter is making. To achieve this contribution, this chapter is set out in five parts. Following Part one (this introduction), Part two will explore justice as meaningful involvement (Hamilton, 2019b, 2021); this conceptualisation of justice draws on Schlosberg's (2007) notions of recognition, participation and capabilities as applied through the lenses of environmental justice, ecological justice, climate justice, species justice and earth jurisprudence/deep ecology. It is an expansive conceptualisation as it draws on multiple notions of justice. While it starts with an expansive notion of justice it was previously narrowed to focus on environmental offending, including in my previous work. Part two will (re)frame that conceptualisation for a legal activities context.

Part three will explore the concept of restorative justice conferencing and explain how it can operate as a vehicle for the achievement of justice as meaningful involvement following environmental offending. The success of conferencing following environmental offending has seen some scholarship exploring its potential in the face of environmental harm from legal activity. Such exploration does not consider what justice as meaningful involvement looks like for legal activity. Part four will rectify that omission, before this chapter is wrapped up with some concluding remarks in Part five.

2 Justice as Meaningful Involvement

As originally formulated, justice as meaningful involvement is an expansive conceptualisation in that it is engaged at the interface of human and environment interaction regardless of the genesis of that interaction. For example, crime, lawful activity and climate change. It entails recognition, participation and capability, the requirements for which are set out in Table 1.

A disadvantage of the expansiveness of this original conceptualisation is that it is not sophisticated enough to draw out the nuances between illegal and legal activities which harm the environment which is at the centre of green criminology. When justice as meaningful involvement was formulated in its expansive framing, a framing in the context of environmental offending was also explored as the dictates of that research required such framing. However, no specific framing in the context of lawful activity was considered. The next section of this chapter will explore justice as meaningful involvement in both an illegal and legal activities context, the purpose being to show that the dictates of justice entail different things depending on the context.

Table 1 Justice as meaningful involvement—as originally conceptualised

Justice as Meaningful Involvement—Original Expansive Conceptualisation (Hamilton, 2019b, 2021)	
Recognition	An acknowledgment that human/environment interaction can have harmful effects on both humans and environment alike (Hamilton, 2021: 236–238)
Participation	The inclusion of a wide range of people, operators, communities, the environment (and its constituent parts) and countries in decisions involving human/environment interaction that may impact on them (Hamilton, 2021: 238–239)
Capability	The ability of those harmed by human/environment interaction to retain (or have restored) that which is essential to their/its functioning (Hamilton, 2021: 239–241)

A. *Recognition (Illegal v Legal Activities)*

Given green criminology's acknowledgment that harm to the environment can be caused by both illegal and legal activities, it is not surprising that the same entities are harmed by both forms of activities—individuals, the environment, communities and commercial operators. Those injured by environmental offending can be referred to as 'victims' (adopting the terminology from mainstream offending), and those injured by legal activity can be referred to as 'those harmed'. These labels reinforce the genesis behind injury, one legal and one illegal. Those harmed by lawful activity are often done so because of the 'ecological-economic trade-off' (Lampkin & Wyatt, 2019) which underpins our Western consumeristic society where some harming of the environment is tolerated in the name of social/economic/technological advance. This means that '[t]he reality of our age is that much of the economy is based on the exploitation of natural resources' (Skinnider, 2013: 3). Indeed, the use of the terminology of 'natural resources' is an anthropocentric positioning which sees humans sitting outside and above nature, with nature being simply a resource to exploit.

Therefore, recognition as the first component of justice as meaningful involvement simply requires the acknowledgment that environmental offending and legal activity involving the environment (collectively, human/environment interaction) can each cause harm to individuals, the environment, communities and commercial operators.

B. *Participation (Illegal v Legal Activities)*

Justice as meaningful involvement requires participation of victims and those harmed, but the participation requirement differs depending on whether the harming behaviour was illegal or legal. In the realm of environmental offending, participation 'entails the involvement of relevant participants (offender, victims, community and stakeholders) in the process used to deal with environmental offending' (Hamilton, 2021: 243). That is, it operates post harm, such as in a prosecution or an environmental restorative justice ('ERJ') conference embedded in a prosecution. Participation as it pertains to lawful activity operates in the face of harm entailing the inclusion of those harmed or potentially harmed by legal activity in decision-making pertaining to the granting of permission to individual activities which may harm the environment or government policy which permits such activity. It dictates the participation of those harmed or potentially harmed by lawful activity in two contexts. Firstly, in decision-making relating to the grant of permission for individual activity which will/may harm the environment. Such participation may be in the form of submissions on a development proposal, feedback into an environmental impact assessment, species impact assessment or cultural impact assessment. Secondly, it dictates participation of those harmed/potentially harmed in government policy formulation permitting such development. This may be in the form of feedback into policy formulation itself, such as government policy around coal extraction and burning. The importance of public participation is reinforced in the Aarhus Convention (UNECE, 1998) and given effect through environmental impact assessment (Lambropoulos, 2010; Macintosh, 2010). The differentiation between participation following environmental offending and participation in the face of legal activity is vitally important when assessing restorative justice as a potential vehicle to achieve justice as meaningful involvement.

In sum, participation as the second component of justice as meaningful involvement requires the participation of relevant stakeholders (offender, victim, helper) in the process used to deal with environmental offending. In a legal activity context, participation is the involvement of those harmed or potentially harmed in decision-making pertaining to permissible activity or the formulation of policy underpinning that permissible activity.

C. *Capability (Illegal v Legal Activities)*

Justice as meaningful involvement requires the retaining or restoring of functioning of those harmed by environmental offending, and the consideration of those harmed or potentially harmed by legal activity to retain (or have restored) that which is essential to functioning. In terms of legal activity, capability of those harmed is not assured (only the consideration of functioning in decision making). This is because the ecological-economic trade-off dictates that there will be occasions where environmental harm is permitted which effectively closes off functioning in the name of social/economic/technological advance. To ensure capability as a component of justice as meaningful involvement is achieved, closing off only occurs after relevant appraisal. That is, after being considered in the decision-making process leading to permission for lawful activities, or the formulation of policy under which the permission for lawful activities is formulated. This is an anthropocentric view of the environment, but it accords with the reality of our society. While an ecocentric view could have been formulated, for example where capability in relation to lawful but awful activity was absolute (that is, had to be ensured) that would not accord with modern reality (for better or worse) (for an overview of capability theory, see Nussbaum, 2006; Sen, 2005). A summary of the requirements of recognition, participation and capability in the context of illegal v legal activity is provided in Table 2.

The differentiation between justice as meaningful involvement in the context of environmental crime vis-à-vis lawful activity is important because it allows an assessment of the mechanisms which can be used to achieve justice in that frame. One such mechanism is ERJ conferencing in the context of environmental offending. Hamilton (2019b, 2021) concluded that ERJ conferencing used as part of the prosecution of environmental offending could achieve justice as meaningful involvement, as it respected recognition, participation and capability. The next part of this chapter will explore ERJ, its application to environmental offending and the achievement of justice as meaningful involvement.

3 Environmental Restorative Justice Following Offending: Achieving Justice as Meaningful Involvement

ERJ is the application of the principles and practices of restorative justice to environmental concerns. There are three main conceptualisations of restorative justice—an ethos; an outcome; and a process. Restorative justice can be viewed as an ethos; a set of principles which underpins

Table 2 Justice as meaningful involvement—as pertains to Illegal v Legal activity

Justice as Meaningful Involvement—Environmental Offending v Lawful Activity		
	Pertaining to environmental offending (Hamilton, 2019b, 2021)	Pertaining to lawful activity (this volume)
Recognition	An acknowledgment that environmental offending can have harmful effects on humans (both currently living and future generations), the environment (and its constituent parts), communities (both Indigenous and non-Indigenous), and commercial operators (Hamilton, 2021: 243)	An acknowledgment that permitted activity, or activity not legally prohibited, which affects the environment, can manifest in harm to individuals (both currently living and future generations), communities (Indigenous and non-Indigenous), the environment (inclusive of its constituents parts including the physical environment, flora, fauna, ecosystems and the biosphere), and commercial operators (collectively 'those harmed')

(continued)

Table 2 (continued)

Justice as Meaningful Involvement—Environmental Offending v Lawful Activity		
	Pertaining to environmental offending (Hamilton, 2019b, 2021)	Pertaining to lawful activity (this volume)
Participation	The involvement of relevant participants (offender, victims, community and stakeholders) in the process used to deal with environmental offending (Hamilton, 2021: 243)	The inclusion of those harmed or potentially harmed by legal activity in decision-making pertaining to the granting of permission to individual activities which may harm the environment or government policy which permits such activity
Capability	The ability of those harmed by environmental offending to retain (or have restored) that which is essential to their/its functioning (Hamilton, 2021: 243)	When undertaking decision making pertaining to the granting of permission to individual activities which may harm the environment or government policy which permits such activity, capability requires the consideration of those harmed or potentially harmed by legal activity to retain (or have restored) that which is essential to functioning

one's life, dictating how one interacts with other people (Gavrielides, 2007; Zehr, 1990/2015a). This ethos dictates that relationships are taken seriously, people acknowledge their impact on the environment and others; responsibility, listening, dialogue and repair are encouraged, decision making is inclusive, conflicts and harms are viewed as opportunities, truths are not imposed, and injustices are confronted (Zehr, 1990/2015a): 95–96).

Restorative justice can also be concerned with outcomes, regardless of the process, and which can be imposed on offenders to repair harm that has been occasioned (McCold, 2000; Zernova & Wright, 2011: 92–93). An example of this outcome focus is the plethora of reparative orders (White, 2017) that the Land and Environment Court of New South Wales (Australia) can hand down when sentencing an environmental offender. One such order can require an offender 'to carry out a specified project for the restoration or enhancement of the environment in a public place or for the public benefit' or pay to have that work undertaken (*Protection of the Environment Operations Act 1997* (NSW): ss 250(1)(c), (e); for an overview of the available orders see Hamilton, 2016).

Another framing of restorative justice is focussed on the encounter it entails. The encounter sees the bringing together, usually face-to-face, of relevant stakeholders to offending (or conflict), including offender, victim, prosecutor (depending on the jurisdiction) and environmental experts, under an independent facilitator. This process/encounter, in the form of a conference, is designed to allow input from and interaction among the stakeholders with a focus on repairing harm, which will manifest in an outcome plan to address the harm occasioned. The restorative justice conference can be held as a form of diversion from prosecution, enabling the offender to avoid prosecution by undertaking a restorative process and committing to, and implementing, an outcome plan. The process is known as the 'front-end model' (see Hamilton, 2022a: 600–601). Statistics around the use of this model are hard to gauge because there do not appear to be public registers of held conferences, and there are no court judgments pertaining to the conferences, unless the conference misfires (see for example, *Canterbury Regional Council v Bathurst Coal Limited* [2019] NZDC 14416). Examples of the front-end model come from New Zealand—Environment Canterbury (for an overview see Hamilton, 2022a: 597–600), West Coast Regional Council (2018) and Environment Southland (2017) and Canada—British Columbia (British Columbia Ministry of Environment, 2012a, 2012b). The back-end model on the other hand embeds the conference as part of the court process (for an overview see Hamilton, 2022a: 597–600). That is, the court considers the fact that the conference occurred and the agreed outcome plan in its approach to sentencing. Most of the use of this model of conferencing comes from New Zealand, where 69 such conferences have been held between 2002 and 2022 (Hamilton & Stanton, 2024).

New South Wales, Australia, has utilised an ERJ conference following environmental offending on two occasions. Those two cases were prosecutions for offending against Aboriginal cultural heritage. I will borrow one of those cases from NSW—*Chief Executive, Office of Environment and Heritage v Clarence Valley Council* (2018) 236 LGERA 291 ('*Clarence Valley Council*'; for an overview see Hamilton, 2019a), and one from New Zealand—*Canterbury Regional Council v Interflow (NZ) Limited* [2015] NZDC 3323 ('*Interflow*'), to highlight ERJ embedded in the prosecution process and how it can achieve justice as meaningful involvement. *Clarence Valley Council* involved the prosecution of the local council after it had cut down a scar tree which had cultural heritage value for the local Aboriginal community. *Interflow* involved the prosecution of a roads contractor which polluted a local stream during road works, causing injury and death to fish and eels and degradation to the stream. This chapter is concerned with restorative justice as an encounter because it has been shown to achieve justice as meaningful involvement (Hamilton, 2021). Recognition of the harm that can flow from environmental offending is embedded in the illegality of the act. That is, the genesis behind the government prohibiting certain harm to the environment is the recognition that such activity can cause harm. The repair of harm, especially when restorative justice is employed as a process, as an encounter using conferencing for example, is a primary driver. Repair of harm is premised on the recognition that such harm is possible in the first place.

Criticisms directed at traditional prosecution encompass limited engagement between victims and offenders, failure to address victims' needs adequately, lack of accountability imposed on offenders for their actions, and impact of stigmatising shame (Braithwaite, 1989).

Christie (1977) and Zehr (1990/2015a, 1990/2015b) have expressed criticism of traditional prosecution's limited inclusion of offenders and victims, which in turn provides no or limited interaction between them. Christie (1977: 9) views offenders and victims as losing their conflict to the state, meaning they have limited opportunity to participate in the resolution of their own conflict; '[t]he offender has lost the opportunity to explain himself [or herself] to a person whose evaluation of him might have mattered [i.e. the victim]'. Zehr (1990/2015b: 37–38) views victims as no more than 'footnotes to the crime…Those who have most suffered are not…part of the resolution of the offence'. Offenders, on the other hand, are treated as bystanders with decisions being made for them. This means that victims need to rely on stereotypes to make sense of what has happened (Christie, 1977: 3, 7, 8), 'offender's misattributions and rationalisations about their offending and victims are not challenged…' (Hamilton, 2021: 82), and crime becomes 'mystified and mythologized until the actual experiences and motivations' behind the crime are lost (Zehr, 1990/2015b: 37–38). In other words, notwithstanding the recognition of harm underpinning prosecution, the prosecution process itself does not translate that recognition into participation.

ERJ responds to those criticisms: it involves input from, and interaction between offender and victim; it meets victims' needs; it makes offenders accountable; and it maximises the use of reintegrative shaming (the opposite of stigmatising shame). Reintegrative shaming can lead to a better relationship between the regulator and those regulated, leading to lower reoffending. There is also a benefit to come from offender and victim interaction, and that is better outcomes which repair harm that has been occasioned (Al-Alosi & Hamilton, 2019: 1467–1472; Hamilton, 2021: 157–175).

A benefit of ERJ over traditional prosecution is the expanded role that offenders and victims play in the process. In fact, they are central. Participation is central to ERJ, especially in the form of a process where conferencing is used. Depending on the case, offenders and victims may be represented by proxies. In cases where offending is by an organisational entity such as a local council, corporation or utilities provider, representation can be through members of those organisations. For example, in *Clarence Valley Council*, where a local council cut down a sacred Aboriginal scar tree, the council was represented at conferencing by the mayor, deputy mayor, general manager and the council officers who removed the tree. In the *Interflow* ERJ conference, the offending corporation was represented by senior management. The participating offender or representatives will be assessed by the independent facilitator for suitability for conferencing (for an overview of suitability criteria see Al-Alosi & Hamilton, 2019, 2021). It is essential that a representative of a non-human offender has knowledge of the offending corporation's activities and culture and is able to bind the organisation to any outcome agreed at conferencing (Hamilton, 2021: 200–201).

Victims of environmental offending are diverse—individuals (e.g. directly affected by pollution, those who have lost a recreational pursuit, those who have lost something of cultural significance, those who have lost a commercial interest), communities, the environment (e.g. trees, plants, animals, ecosystems, rivers) and commercial operators (Hamilton, 2019b: 214). Not all these victims can speak in the way that you and I do, hence representatives are needed to represent them at conferencing. A recent review of the use of ERJ conferencing in New Zealand reveals a wide array of victim representatives—'the local Māori Trust, chairperson of a local community board and walkway trust, community organisations', 'local Indigenous people and groups who have a close connection with the environment' (Hamilton & Stanton, 2024: 603–604; citations omitted). In Australia, when an ERJ conference was used following offending against Aboriginal cultural heritage, members of the Local Aboriginal Land Council attended as victims and victim representatives (for an overview see Hamilton, 2008, 2019a). In *Interflow*, the local Māori people (the Ōuku Rūnanga) represented Walnut Stream (including its inhabitants—fish and eel etc.) and its connection with the people; they expressed 'how the wellbeing of the stream directly impacted on the mauri [life force or essence] and people of the land' (*Interflow*: [18]).

The flow on effect of limited offender and victim engagement and input in traditional prosecution is that victim needs are not met (Zehr, 1990/2015b: 39, 67). Four victim needs can be considered—information, truth-telling, empowerment and restitution/vindication. While these may not be the only victim needs, they seem to be the main ones. It is the meeting of these needs which is essential for achieving capability, as a component of justice as meaningful involvement. The information surrounding offending is often held by the offender including motives which may not always come out during a court case. Being aware of such information may help a victim understand the offending better and be able to predict the likelihood of such an offence happening again in the future. In an environmental offending context, the victim may want to know *why you (Joe Blow or XYZ Corporation) dumped hazardous pollution in a local swimming hole resulting in injury to persons who swim in the river and to the river itself*. As the conference is a face-to-face dialogue, the offender will be asked such questions and may give the answer *it was a financial decision as I/my company is facing increasing fees to dispose of such waste*. The conversation could then move on to meet another victim need, truth-telling—'the ability to tell their story: what happened and how it has affected them and their loved ones' (Hamilton, 2021: 83). This may take the following form, where a victim may say to an offender: *that is very selfish and irresponsible of you. I spent a week in hospital because of pollution sickness and had to miss work. Now the local authorities advise that we can't swim in the river for another two months until your mess is cleaned up. I am upset and very disappointed*.

Another victim need is empowerment. Empowerment can come from a victim taking part in a conference, seeking the information they need, telling their truth, and coming up with ways to repair the harm that has been caused. Another is restitution/vindication. Restitution as a form of compensation may act as a form of vindication for the victim—an acknowledgement that they have done nothing wrong. In our exchange, it may be that the offender may want to compensate the swimmer for her medical bills and time off work, or otherwise as compensation for pain and suffering or loss of a recreational pursuit.

Holding offenders accountable is another benefit of ERJ conferencing. Traditional accountability following offending is quite narrow—'the laying of charges, the prosecution of those charges and the punishment of an offender when they or it are found guilty of the charge(s) or plead guilty to those charges' (Hamilton, 2021: 68). Conferencing achieves a wider form of accountability involving an offender appreciating the harm they have caused, facing up to what they have done, and taking steps to repair the harm (Zehr, 1990/2015a: 24, 1990/2015b: 47). It is the direct offender and victim input and interaction which conferencing facilitates which leads to accountability in this wider sense.

Restorative justice conferencing allows for the deployment of reintegrative shame as opposed to stigmatising shame. Stigmatising shame, the product of traditional prosecution, sees shame directed at the person—*Johnny you committed that crime because you are a bad person and will never amount to anything*. Expanding Braithwaite's (1989) conceptualisation of stigmatising shame to environmental offending—*XYZ Corporation you are a renegade operation and a scourge and will need constant supervision.* Reintegrative shaming on the other hand is conducive to reintegration and closer regulator/regulated relationships. For mainstream offending—*Johnny you have done a bad thing, but you are a good person who made a bad choice. We will work with you to help you through this experience.* In an environmental offending context—*XYZ Corporation you have caused harm through your operations. The regulator will work with you so that you will operate as a responsible corporate citizen.* In an environmental regulatory context, the deployment of reintegrative shaming during a conference can strengthen the relationship between an offending organisation and the regulatory authority through the taking of a cooperative problem-solving approach which attempts to repair harm from the current offending but also prevent future harm through operational and procedural improvements to the offending organisation.

The input from, and interaction between, offender and victim at conferencing (achieving participation) can lead to better outcomes (to ensure capability) than traditional prosecution because victims are asked what they need and offenders are asked what they can do to repair the harm that has been occasioned. While environmental prosecution has traditionally resulted in a fine (at least in NSW; see Hamilton, 2019b: Figs 8 and 9), this has given way to environmental repair type orders, such as environmental service orders (see Hamilton, 2019b: Fig 10).

While this pivot is desirable, it is limited without victim, offender and expert input because the repair may not target the actual harm occasioned but some other aspect of the environment. This was found to be the case with sentencing of two forms of environmental offending in NSW, where the majority of the environmental service orders made had a weak nexus to the harm that had been occasioned by the offending (Hamilton, 2019b; Figs 12 and 13; see Table 8 for the definition of a 'weak nexus').

Capability was assured in the *Interflow* proceedings through an outcome devised at conferencing where an $80,000 donation was made to a community group (Banks Peninsula Conservation Trust) to undertake the repair/enhancement of the stream, thereby allowing the stream to function, eels and fish to live, and the local Māori people to enjoy their connection with the stream and its contribution to their life force (*Interflow*: [43]). In *Clarence Valley Council*, capability was about ensuring the protection of Aboriginal heritage and the significant stories which attach to that heritage. The outcomes reached at the conference are illuminating of that point. Firstly, cultural awareness training of council staff, so that staff are aware of ways Aboriginal people express and preserve their culture, through for example, scar trees. Secondly, a project meant to ensure 'that the scar tree will live on, albeit as remnants and perhaps in sculpture form, its history preserved, and the site commemorated' (Hamilton, 2019a: 208). Enabling such a project was a $300,000 payment from the council to the Grafton Ngerrie Local Aboriginal Land Council for a tree restoration and interpretation project directly related to the scar tree (*Clarence Valley Council*: [19]).

Notwithstanding the benefits of ERJ, limitations have been raised. However, these limitations spring from a misunderstanding of ERJ, and are therefore misconceived and can be overcome. A limitation of conferencing for some people, particularly those steeped in traditional prosecution, is that it is not adversarial. That is true, but what must be remembered is that conferencing does not determine innocence or guilt and therefore does not need to be adversarial. Indeed, if the conference were adversarial it would detract from what it is designed for—an interaction between stakeholders to an offence (see further Hamilton, 2021: 175).

Some people conceive conferencing as a soft option for offenders. These views are more likely to be held by those steeped in the adversarial system and those that see punitiveness as part of the prosecution process. What this fails to consider is the emotional and cognitive load an offender must carry when coming face-to-face with people they have harmed through offending. An offender listening to a victim's truth-telling and responding to their need for information is likely to be a lot more demanding than hiding behind their lawyer in court. Those that adhere to this soft option idea must also recognise that any outcome plans are considered in sentencing, with sentencing still the domain of the sentencing judge. The judge may ignore the outcome plan, adopt it, change it and sentence in addition to it (see further Hamilton, 2021: 176–177).

Consistency and proportionality in sentencing may be affected by conferencing, usually in a situation where an outcome plan results in a harsher penalty than a court would have ordered. This was the case in *Interflow*, where the offender agreed at conferencing to a donation of $80,000 to repair the harm it had caused to a stream despite knowing that the likely fine imposed by the court would be $40,000. On its face, this $80,000 donation is not proportionate to the offending committed, notwithstanding that it is needed to repair the stream, and inconsistent with similar cases that proceed through court and a fine of $40,000 is ordered. An apples and oranges comparison can be made here. That is, court proceedings (apples) should not be compared with court proceedings that utilise conferencing (oranges) because they are different things. There may be benefits to an offender which justify the extra expenditure of money, such as making amends, repairing harm, repairing relationships, repairing or affirming a social licence to operate (for an overview of social licence see Bice, 2014; Luke, 2018), as part of corporate social responsibility (for an overview see Aguinis et al., 2024), or to repair reputation (see further Hamilton, 2021: 177–178; 2022a: 604–606).

One final limitation of conferencing is the investment of time, money and effort needed. That may be true, someone needs to pay for the independent facilitator's time preparing for the conference, conducting the conference, and in any follow-up following the conference. Usually, that expense is met by the offender and should be considered in sentencing.

In terms of time, conferencing takes time because of the preparation and holding of the conference. This may interfere with a court's time matrices where time to complete matters is used as an indicator for performance. The easiest way to overcome this limitation is to separate matters which use conferencing from those that don't use conferencing. Again, it goes back to the apples and oranges argument I made earlier. Conferencing does require more effort by an offender and victim than traditional prosecution, but that is not a limitation in itself because of the benefits which can derive from that investment of effort (Hamilton, 2021: 178). Having explored ERJ conferencing in the context of environmental offending, and how it operates as a vehicle to achieve justice as meaningful involvement, this chapter will move on to consider the framing of justice as meaningful involvement in the context of legal activity and consider whether ERJ has a role in achieving justice in that framing.

4 Conceptualising Justice in the Context of Lawful Activity: Justice as Meaningful Involvement

Spurred on by the apparent success of ERJ conferencing as a vehicle to achieve justice as meaningful involvement, it is tempting to simply transplant that application to lawful activities which harm the environment. Indeed, those affected by environmental offending and lawful activities are the same—individuals, communities, the environment, recognising that both illegal and legal activities can have harmful effects. It is participation and capability which differentiate what justice as meaningful involvement entails for illegal versus legal activity, and it is that differentiation which must be considered when exploring ERJ.

As explored earlier in this chapter, ERJ conferencing in the context of environmental offending brings together various stakeholders (offender, victim and helpers) (participation) to talk about the harm that has been occasioned (recognising the harm environmental offending can cause) and then work to find ways to repair that harm (i.e. retain or have restored functioning—capability). The conference could be a diversion from prosecution, or it could be embedded in the prosecution process. Either way, it is post offending and fits clearly within participation in a 'process used to deal with environmental offending' (Hamilton, 2021: 243).

My earlier research explored the possibility of an ERJ conference in the context of lawful activity which harmed, or had the potential to harm, the environment (Hamilton, 2022b). Regrettably, that exploration occurred without justice as meaningful involvement having been formulated and therefore considered in that specific context. What this means is that the focus of the analysis was misguided. Akin to an ERJ conference occurring in the context of environmental offending, it was proposed the entity causing or potentially causing the environmental harm would meet with those harmed or possibly harmed in one of two stages:

- Pre-operations, '[a]t the application, assessment, approval and implementation stage..., as a means of delineating the expected environmental harm and explaining why such environmental harms are unavoidable, as a forum to explore some of the benefits to bestow on the local community through the activity or project, and, to establish a social licence to operate' (Hamilton, 2022b: 380).
- Operations 'stage when environmental harm is lawfully caused by the activity..., as a means of maintaining and reinforcing that which was established during the ...[pre-operations stage conferencing]: dialogue, relationships and social licence to operate...' (Hamilton, 2022b: 380).

The use of ERJ conferencing in a legal harm context does not meet the *participation* requirement in the conceptualisation of meaningful involvement, as that entails those potentially harmed feeding into the decision-making process pertaining to the lawful activity or the policy which underpins the lawfulness of such activity. Ultimately, those decisions are made by the government following public input in the form of submissions, relating for example, to environmental impact assessment, species impact assessment and cultural impact assessment. Therefore, an ERJ conference involving victims and offenders, would not be achieving participation. Even if the government decision maker did attend the ERJ conference, it may prove difficult to hear from everyone who wanted to comment. It seems more appropriate that the decision maker have the opportunity to consider the written submissions in a reasonable time, which does not need a conference to achieve. That is, to suggest an ERJ conference as a vehicle for decision making around lawful activity (either policy formulation or pertaining to individual projects) seems inappropriate and misguided.

The use of ERJ conferencing in a legal harm context does not meet the capability requirement in the conceptualisation of meaningful involvement, because what that entails is the decision maker considering the functioning of those potentially harmed by its decision. The ERJ conferences proposed in the earlier research did not contemplate the presence of the decision maker, only the proponent (Hamilton, 2022b), and thereby placing some onus on the proponent to make good any harm they have caused legally. I ask rhetorically, what would be the motivation of a proponent of lawful activity to attend a conference and make good any harm they have caused legally (absent a legal requirement to do so) where no legal (and arguably no moral) blameworthiness attaches to their activities? Notwithstanding the fact that some proponents may welcome such a conference to, for example, strengthen their social licence to operate or because they are a good corporate citizen, that does not displace the fact that it is the responsibility of the decision maker to consider capability and not the proponent. That consideration comes after public submissions are made (participation) and does not require a conference. Hence, for the achievement of participation and capability, what is needed is a robust public consultation process which entails submissions on proposed developments and government policy. This accords with 'a commonly held view that public participation in environmental decision-making results in better environmental outcomes' (McNamara & Crane, 2018: 320). Public participation can be achieved without the need for a conference.

5 Conclusion

Justice as meaningful involvement, encompassing recognition, participation and capability, can be achieved in the context of environmental crime where ERJ is employed (Hamilton, 2021). While this is important, it is limited when considering the totality of environmental harm/damage/destruction for three reasons. Firstly, environmental crime only accounts for a portion of total environmental harm/damage/destruction. Indeed, much harm comes from permitted activity. Secondly, only a portion of environmental crime is prosecuted, and thirdly, only a portion of environmental prosecution will engage with a conference.

Given that a substantial portion of total environmental harm/damage/destruction comes from government decision making (through approval of consent, licence and permits, and policy development), it is important to conceive of justice as meaningful involvement in the context of lawful activity. This chapter has framed that conceptualisation. Suggesting the use of ERJ in the context of lawful activity without having a conceptualisation of justice as meaningful involvement in mind is misguided. As analysis in this chapter has shown, both participation and capability have different requirements for illegal activity compared to legal activities which limits the usefulness of ERJ in the context of legal activity.

References

Aguinis, H., Rupp, D. E., & Glavas, A. (2024). Corporate Social Responsibility and Individual Behaviour. *Nature Human Behaviour, 8*, 219–227. https://doi.org/10.1038/s41562-023-01802-7

Al-Alosi, H., & Hamilton, M. (2019). The Ingredients of Success for Effective Restorative Justice Conferencing in an Environmental Offending Context. *University of New South Wales Law Journal, 42*(4), 1460–1492. https://doi.org/10.2139/ssrn.3495322

Al-Alosi, H., & Hamilton, M. (2021). The Potential of Restorative Justice in Promoting Environmental Offenders' Acceptance of Responsibility. *University of New South Wales Law Journal, 44*(2), 487–512. https://doi.org/10.53637/GBEQ9308

Bice, S. (2014). What Gives You a Social Licence? An Exploration of the Social Licence to Operate in the Australian Mining Industry. *Resources, 3*(1), 62–80. https://doi.org/10.3390/resources3010062

Braithwaite, J. (1989). *Crime, Shame and Reintegration.* Cambridge University Press.

British Columbia Ministry of Environment. (2012a). *Community Environmental Justice Forums Policy.*

British Columbia Ministry of Environment. (2012b). *Community Environmental Justice Forums.*

Canterbury Regional Council v Bathurst Coal Limited [2019] NZDC 14416.

Canterbury Regional Council v Interflow (NZ) Limited [2015] NZDC 3323.

Celermajer, D., Schlosberg, D., Rickards, L., Stewart-Harawira, M., Thaler, M., Tschakert, P., Verlie, B., & Winter, C. (2021). Multispecies Justice: Theories, Challenges, and a Research Agenda for Environmental Politics. *Environmental Politics, 30*(1–2), 119–140.

Chief Executive, Office of Environment and Heritage v Clarence Valley Council (2018) 236 LGERA 291.

Christie, N. (1977). Conflicts as Property. *British Journal of Criminology, 17*(1), 1–15.

Environment Southland. (2017). *Compliance Policies.*

Gavrielides, T. (2007). *Restorative Justice Theory and Practice: Addressing the Discrepancy*. European Institute for Crime Prevention and Control.

Goyes, D. R., & Sollund, R. (2018). Animal Abuse, Biotechnology and Species Justice. *Theoretical Criminology, 22*(3), 363–383. https://doi.org/10.1177/1362480618787179

Hamilton, M. (2008). Restorative Justice Intervention in an Environmental Law Context: Garrett v Williams, Prosecutions under the Resource Management Act 1991 (NZ), and Beyond. *Environmental and Planning Law Journal, 25*(4), 263–271.

Hamilton, M. (2016). Restorative Justice Intervention in an Environmental and Planning Law Context: Applicability to Civil Enforcement Proceedings. *Environmental and Planning Law Journal, 33*(5), 487–501. https://doi.org/20.500.14802/6917

Hamilton, M. (2019a). Restorative Justice Intervention in an Aboriginal Cultural Heritage Protection Context: Chief Executive, Office of Environment and Heritage v Clarence Valley Council. *Environmental and Planning Law Journal, 36*(3), 197–211. https://doi.org/20.500.14802/6889

Hamilton, M. (2019b). *Restorative Justice Conferencing in Response to Pollution Offending: A Vehicle for the Achievement of Justice as Meaningful Involvement* (PhD Dissertation). UNSW.

Hamilton, M. (2021). *Environmental Crime and Restorative Justice: Justice as Meaningful Involvement*. Palgrave Macmillan. https://doi.org/10.1007/978-3-030-69052-6

Hamilton, M. (2022a). Restorative Justice Conferencing in a New Zealand Environmental Offending Context: Two Models. In B. Pali, M. Forsyth, & F. Tepper (Eds.), *The Palgrave Handbook of Environmental Restorative Justice* (pp. 593–616). Palgrave Macmillan. https://doi.org/10.1007/978-3-031-04223-2_23

Hamilton, M. (2022b). Restorative Justice Conferencing: A Vehicle for Repairing Harm Emanating from Lawful But Awful Activity. In J. Gacek & R. Jochelson (Eds.), *Green Criminology and the Law* (pp. 361–386). Palgrave Macmillan. https://doi.org/10.1007/978-3-030-82412-9_16

Hamilton, M., & Stanton, E. (2024). Characteristics of Restorative Justice Conferencing in a New Zealand Environmental Offending Context. *New Zealand Law Review, 4*, 585–616.

Lambropoulos, V. (2010). What Can Australia Learn from the Europeans about Public Participation? Article 6 of the Aarhus Convention and Environmental Impact Statements. *Environmental and Planning Law Journal, 27*(4), 272–295.

Lampkin, J., & Wyatt, T. (2019). Utilising Principles of Earth Jurisprudence to Prevent Environmental Harm: Applying a Case Study of Unconventional Hydraulic Fracturing for Shale Gas in the United Kingdom. *Critical Criminology*. https://doi.org/10.1007/s10612-018-9426-7

Luke, H. (2018). Not Getting a Social Licence to Operate Can Be a Costly Mistake, as Coal Seam Gas Firms Have Found. *The Conversation.*

Macintosh, A. (2010). Best Practice Environmental Impact Assessment: A Model Framework for Australia. *Australian Journal of Public Administration, 69*(4), 401–417. https://doi.org/10.1111/j.1467-8500.2010.00703.x

McCold, P. (2000). Toward a Holistic Vision of Restorative Juvenile Justice: A Reply to the Maximalist Model. *Contemporary Justice Review, 3*(4), 357–414.

McNamara, N., & Crane, W. (2018). Public Participation and the Adani Syndrome. *Environmental and Planning Law Journal, 35*(3), 320–330. https://search.informit.org/doi/10.3316/agispt.20182299

Murdock, E. G. (2021). A History of Environmental Justice. In B. Coolsaet (Ed.), *Environmental Justice: Key Issues* (1st ed.). Routledge.

Nussbaum, M. C. (2006). *Frontiers of Justice: Disability, Nationality, Species Membership*. Harvard University Press.

Porter, L., Rickards, L., Verlie, B., Bosomworth, K., Moloney, S., Latham, B., Anguelovski, I., & Pellow, D. (2020). Climate Justice in a Climate Changed World. *Planning Theory & Practice, 21*(2), 293–321. https://doi.org/10.1080/14649357.2020.1748959

Protection of the Environment Operations Act 1997 (NSW).

Schlosberg, D. (2007). *Defining Environmental Justice: Theories, Movements, and Nature*. Oxford University Press.

Sen, A. (2005). Human Rights and Capabilities. *Journal of Human Development, 6*(2), 151–166. https://doi.org/10.1080/14649880500120491

Skinnider, E. (2013). *Effect, Issues and Challenges for Victims of Crimes That Have a Significant Impact on the Environment*. International Centre for Criminal Law Reform and Criminal Justice Policy.

United Nations Economic Commission for Europe (UNECE) (1998). *Convention on Access to Information, Public Participation in Decision-Making and Access to Justice in Environmental Matters*. Aarhus, Denmark.

West Coast Regional Council. (2018). *Compliance and Enforcement Policy*.

White, R. (2008). *Crimes Against Nature: Environmental Criminology and Ecological Justice*. Willan.

White, R. (2017). Reparative Justice, Environmental Crime and Penalties for the Powerful. *Crime, Law and Social Change, 67*, 117–132. https://doi.org/10.1007/s10611-016-9635-5

White, R. (2018). Green Victimology and Non-Human Victims. *International Review of Victimology, 24*(2), 239–255. https://doi.org/10.1177/0269758017745615

Zehr, H. (1990/2015a). *The Little Book of Restorative Justice*. Good Books.

Zehr, H. (1990/2015b). *Changing Lenses: Restorative Justice for Our Times*. Herald Press.

Zernova, M., & Wright, M. (2011). Alternative Visions of Restorative Justice. In G. Johnstone & D. W. Van Ness (Eds.), *Handbook of Restorative Justice* (pp. 91–108). Routledge.

7

Paws with Thought: Legal Recognition of Animal Sentience in the Australian Capital Territory and Beyond

Amber McKinley, Astrid Vachette, and Erica Honey

1 Introduction

Sentience is defined as the capacity to have subjective experiences and feelings, including both negative states, such as pain and suffering and positive states, such as comfort and pleasure (Birch et al., 2021).[1] While animal welfare legislation in Australia has long criminalised cruelty based on the capacity of animals to suffer, such recognition was largely implicit,

A. McKinley (✉) · A. Vachette
Charles Sturt University, Bathurst, Australia
e-mail: amckinley@csu.edu.au

A. Vachette
e-mail: avachette@csu.edu.au

E. Honey
Murdoch University, Murdoch, Australia
e-mail: info@ericahoneyconsulting.com

[1] The first author used Perplexity AI to generate suggestions for reference list and APA7 formatting. They reviewed and edited all AI-generated references to ensure accuracy and appropriateness for this chapter.

R. Killean et al. (eds.), *Conceptualising Justice*, Palgrave Studies in Green Criminology, https://doi.org/10.1007/978-3-032-19791-7_7

with laws focusing on preventing unnecessary harm rather than affirming animals' intrinsic value or emotional lives. The shift towards explicit legal recognition of animal sentience, as seen in the 2019 amendment to the *Animal Welfare Act 1992* (ACT) ('*Animal Welfare Act*'), marks a significant evolution in legal doctrine. This explicit recognition is informed by scientific models such as the Five Domains animal welfare framework, which assesses the impact of nutrition, environment, health, and behaviour on animals' mental state to determine overall welfare (Mellor & Reid, 1994). This framework extends legal duties beyond merely preventing cruelty to promoting positive welfare states and quality of life, such as comfort, interest, and pleasure for all domestic animals.

The 2019 amendments to the *Animal Welfare Act* promote animal sentience as an object of the Act, recognising that 'animals are sentient beings that are able to subjectively feel and perceive the world around them' (*Animal Welfare Act*: s4A(1)(a)). Animals are understood to have the capacity for sensations, feelings, emotions, and experiences that matter to them, including both physical and mental states such as happiness, suffering, and pain (Kotzmann, 2023). The amended law recognises that:

- Animals[2] are sentient.
- Animals have intrinsic value and deserve to be treated with compassion and to enjoy a quality of life that reflects this value.
- People have a duty to care for both the physical and mental welfare of animals (Evans, 2019; Roberts, 2019)

The law applies broadly to domestic animals, livestock, and wildlife, reflecting a community expectation for compassionate treatment and higher welfare standards. This dual acknowledgement of sentience (*Animal Welfare Act*: s 4A(1)) and intrinsic value (*Animal Welfare Act*: s 4A(1)(b)) challenges traditional property-based frameworks under

[2] An animal is legally defined as a living vertebrate species and includes amphibian, fish, bird, a non-human mammal, or a cephalopod e.g. octopus or live crustacean intended for human consumption (*Animal Welfare Act:* Dictionary). For this chapter, 'animal' will refer to vertebrate species (Wilkins, 2019). For an overview of invertebrate justice, see Durrant within this volume.

statutes like the *Crimes Act* 1900 (ACT) ('Crimes Act'), positioning animals as subjects of moral concern rather than objects of ownership. Central to this reform is s 4A(1)(c) of the *Animal Welfare Act*, which imposes a duty to safeguard both physical and mental welfare, reflecting advancements in animal welfare science.

This chapter critically examines the *Animal Welfare Act* and other nations' sentience laws through green criminology's eco-justice framework (see White's chapter, this volume), assessing its role in challenging legal constructions of animals as property and advancing species justice. By analysing the Act's recognition of animal sentience and mental welfare obligations, the chapter interrogates how this reform disrupts anthropocentric legal paradigms under statutes like the *Crimes Act*, while exposing tensions between progressive welfare provisions and enduring exemptions for industries. It evaluates the Act's symbolic and practical contributions to reconceptualising animal victimhood, legal personhood, and duty-of-care norms, while critiquing its limitations in addressing structural speciesism. Situating the Australian Capital Territory ('ACT') reforms within broader ecological justice debates (Schlosberg et al., 2025), the analysis underscores the transformative potential-and constraints-of embedding sentience in law, offering insights into green criminology's evolving engagement with species justice and regulatory reform.

The following sections will discuss the intrinsic value of animals, green criminology's focus on sentience and then discuss animal sentience within an eco-justice framework. The next portion of the chapter will present information related to the challenges to legal recognition of sentience. It will then analyse the significance of legislating for sentience. A conclusion with recommendations will finalise this chapter.

2 Animals' Sentience and Intrinsic Value

Animal welfare can be conceptualised in three ways. First, it can be considered by assessing only the animal's 'biological fitness' e.g. dog breed conformation. Second, welfare can be conceptualised in terms of 'naturalness', for example, can a caged chicken kept for egg production experience natural behaviours such as dustbathing? Third and finally, the 'affective state' addresses the feelings of animals (Broom, 2014; Fraser, 2008; Willis et al., 2021). All three concepts of biological fitness, naturalness and affective state are interrelated and valid when considering animal welfare (Beaver & Golab, 2023).

Contemporary animal welfare standards have evolved to emphasise not only the minimisation of suffering but also the active promotion of positive experiences, particularly mental enrichment and psychological well-being. Central to this shift is the Five Domains Framework, which underscores the importance of providing animals with opportunities for rewarding experiences such as social interaction, play, exploration, and meaningful choice, thereby fostering confidence, resilience, and overall flourishing (Mellor & Reid, 1994). The Five Domains Framework evolved from the World Organisation of Animal Health-endorsed 'Five Freedoms Welfare Framework' which provides that farm animals should have:

1. Freedom from hunger and thirst: by providing ready access to fresh water and a diet to maintain full health and vigour.
2. Freedom from discomfort: by providing an appropriate environment, including shelter and a comfortable resting area.
3. Freedom from pain, injury, or disease: by prevention or rapid diagnosis and treatment.
4. Freedom to express normal behaviour: by providing sufficient space, proper facilities, and company of the animal's own kind.
5. Freedom from fear and distress: by ensuring conditions and treatment that avoid mental suffering (WOAH, 2025).

In 2004, the European research project 'Welfare Quality Project' adapted the 'Five Freedoms' into twelve independent criteria to address more dimensions of animal welfare (Broom, 2022; Canali & Keeling, 2009). The twelve criteria were then organised into four domains of animal welfare. All domains, nutrition, health, physical environment, and behavioural interactions contribute to the mental state of animals (Blokhuis & Veissier, 2023: 25–27). On this basis, Mellor and Beauseil (2015) extended the Five Domains Framework, which considers a fifth dimension, the mental state, and encourages a positive state of beings (Blokhuis & Veissier, 2023: 25–27; Mellor, 2016).

Modern advances in animal welfare science demonstrate that animal welfare can be measured. Contemporary tools to measure animal stress, pain, and discomfort can be used, such as a 'Global Stress Assessment', 'Fear Anxiety Scale' (FAS) in dogs, cats, horses, birds and rats and the species-specific tools such as 'Grimace Scales' (Fear Free, 2025; Fischer-Tenhagen et al., 2022; Pierrette et al., 2023; Uhrig & Robinson, 2006). The measurement of animal welfare can be subjective, as interpretation can depend on the skill, knowledge and ability of human individuals and the development of animal welfare science (Hampton et al., 2023). However, behavioural and environmental enrichment, such as the provision of dynamic environments, cognitive challenges, and social opportunities, are now recognised as essential components of good welfare, enabling animals to express natural, motivated behaviours and maintain good mental health (Taylor et al., 2023). This approach moves beyond the historic focus on alleviating negative states, acknowledging that the absence of physical suffering does not equate to a 'good life'; rather, welfare is optimised when animals are provided with environments and management practices that encourage positive affective states and support their psychological needs (Mellor & Reid, 1994). As a result, welfare frameworks increasingly advocate for the systematic assessment and provision of positive experiences, recognising their critical role in the overall quality of life for animals in human care (Mellor & Reid, 1994; Rault et al., 2025).

Building on this emphasis on animal well-being, scholarly fields such as green criminology have emerged to critically interrogate not only the welfare of animals in human care but also the broader social, legal, and environmental contexts in which animals exist, highlighting the interconnectedness of animal welfare and wider issues of justice and harm, and the rights of animals to 'expect' a level of care and comfort. This is explored further in the following section.

3 Sentience in Green Criminology

Green criminology examines environmental crimes and harms that impact humans, animals, ecosystems, and the biosphere, expanding the notion of victimhood to include not only people but also non-human species and nature itself (Brisman, 2014; Johnson, 2017). Green criminology offers significant analytical depth to global discourse on sentient animal acts by broadening the conversation beyond legal compliance and anthropocentric welfare concerns to encompass questions of harm, ecological justice, and the recognition of non-human animals as victims within multispecies justice frameworks (McEwan & Turley, 2022). Green criminology critiques the limitations of traditional animal welfare law, drawing attention to the societal, cultural, and structural roots of animal harm, as well as how laws and criminal justice institutions conceptualise and address these harms (Duffy et al., 2025).

Applying this lens to the legal recognition of animal sentience situates the ACT legislative reform within green criminology's eco-justice framework (White, 2016) and engages with ongoing debates about species justice (Celermajer et al., 2021) and the recognition of animals as victims rather than objects. Recognising animal sentience within a green criminology holistic approach—integrating criminal, civil, and administrative responses—enables more effective interventions against harms such as wildlife trafficking, as demonstrated by interdisciplinary collaborations in the Global South (McEwan & Turley, 2022).

By foregrounding the sentience and agency of animals, green criminologists advocate for regulatory frameworks and justice mechanisms that address the root causes of harm, including social, economic, and cultural drivers, rather than relying solely on punitive measures (Gibson, 2022; McEwan & Turley, 2022). A green criminological approach also foregrounds the work of Non-Governmental Organisations (NGOs) and community groups in education, advocacy, and law reform, further amplifying positive outcomes for both animals and human communities that depend on healthy ecosystems (Gibson, 2022; McEwan & Turley, 2022). A natural progression from these collaborative efforts is the adoption of broader conceptual frameworks that address the interconnectedness of environmental and social concerns, such as eco-justice frameworks.

Building on Gacek and Jochelson's (2020) argument that the intersection of green criminology and law exposes anthropocentric logics and reconstitutes animals as more than property, the ACT Act can be analysed not just as a reflection of human ethical progression, but as part of wider systems of power, species hierarchies, and societal values that shape which animals receive protection and how infractions are policed and punished. Additionally, green criminological scholarship calls for more than prohibition of cruelty; it seeks to develop policy and legal perspectives that enable ecological and species justice, highlighting the responsibility of justice systems to acknowledge the subjectivity, capacities, and victimhood of sentient animals (McEwan & Turley, 2022).

An eco-justice framework operationalises green criminology's interdisciplinarity by linking animal protection to broader concerns of ecological sustainability and social equity, advancing a conception of justice that extends across human communities, non-human species, and future generations. Integrating ecological sustainability with social fairness, it emphasises equitable distribution of environmental benefits and burdens while challenging anthropocentric systems through recognition of the intrinsic value of non-human life and ecosystems (White, 2013, 2018). Eco-justice is enacted through distributive, procedural, and recognition dimensions, a tripartite model crystallised by Franks and Booker (2018) to address entrenched power imbalances in environmental governance. Scholarship has since expanded its scope: Martusewicz et al. (2015) applied eco-justice to education, advocating the revitalisation of 'cultural commons' to resist consumerist paradigms, while Franks and Booker (2018) demonstrated its application in conservation, particularly in recognising Indigenous stewardship and ensuring procedural equity in resource management. Together, these contributions show how eco-justice provides a critical framework for bridging environmental and social crises by exposing and contesting environmental racism, systemic inequality, and species exploitation.

Species justice, when viewed alongside eco-justice, provides a nuanced framework for understanding animal sentience as both a moral and legal imperative, recognising the need for equitable consideration across species and ecological systems. Eco-justice emphasises not only the interconnectedness of all living beings but also the right of non-human entities to moral standing, and the responsibility to protect future generations and ecosystem integrity. Dentith and Thompson (2017) support a shift towards eco-centric worldviews, advocating for educational and legal systems that recognise non-human entities as deserving of moral standing, which complements species justice's call for interspecies equity and ethical stewardship. Together, these analyses point to the transformative potential of reframing policy and pedagogy towards intergenerational and interspecies justice, urging society to move beyond anthropocentric approaches and embrace frameworks where both humans and non-human animals are considered recipients of justice whose vulnerabilities and needs must be accounted for.

4 Legislating Sentience in the ACT and Beyond

The Australian Royal Society for the Prevention of Cruelty to Animals (RSPCA Australia) endorsed the ACT *Animal Welfare Act's* provisions as aligning with contemporary community expectations, emphasising their role in modernising welfare standards and enforcement mechanisms (RSPCA Australia, 2019). Collectively, these provisions mark a paradigm shift towards species justice, though their transformative potential remains contingent on overcoming enduring regulatory and cultural barriers.

By embedding sentience in the legal framework, the ACT's *Animal Welfare Act* marks a significant shift away from traditional views of animals solely as property or commodities, instead recognising their intrinsic capacity for mental welfare. The Act prohibits practices such as prolonged confinement, reflecting a growing understanding of the complex needs of animals beyond mere survival (*Animal Welfare Act*: Section 9—Unlawful confinement of animal). The introduction of specific offences, such as the failure to exercise dogs, and the empowerment of authorities to impose interim bans, seize, and rehome animals in cases of abuse, reflect a more robust enforcement regime. These measures collectively redefine animals as victims of harm with enforceable rights, rather than as objects whose value is determined solely by human interests. While the legislation maintains certain exemptions for activities conducted under approved codes of practice (Englefield et al., 2019), its explicit sentience clause has symbolic and practical implications, setting a precedent for future animal welfare reforms across Australia (Kotzmann, 2020). Importantly, the Act reframes welfare as an affirmative obligation rather than mere cruelty prevention (Kotzmann, 2023).

Beyond the ACT, a growing number of countries have enacted laws that formally recognise animals as sentient beings. New Zealand amended the *Animal Welfare Act 1999* (NZ) in 2015 to recognise animal sentience, including certain invertebrates such as octopuses and lobsters (*Animal Welfare Amendment Act (No 2) 2015* (NZ); Bionet NZ., n.d.). Across Europe, as of late 2019, at least 30 countries[3] have incorporated animal sentience into their legal frameworks. Further, the European Union itself recognises animal sentience in its foundational treaties (EUR-Lex., n.d.). In North America, the Canadian province of Quebec and several USA states and districts, such as Oregon, Vermont, Maine, and the District of Columbia, have passed laws or amendments acknowledging animal sentience (Animal Legal Defense Fund, 2024; Kotzmann & Stonebridge, 2021; Lessard, 2023). In South America, there is legislation recognising animal sentience in Chile, Colombia, and Peru (Contreras, 2016). In Africa, Tanzania is among the countries with such legal provisions (Bahari et al., 2024).

Recent literature on legislative reform debates reveals a dynamic and contested field shaped by shifting societal values, advocacy, and the limitations of traditional legal frameworks. Scholars such as Kotzmann (2020), highlight the significance of legislative moves like the ACT's explicit recognition of animal sentience, noting how parliamentary debates framed this reform as both symbolically and practically transformative in redefining animals as beings with intrinsic value and

[3] Including: Austria, Belgium, Bulgaria, Croatia, Cyprus, the Czech Republic, Denmark, Estonia, Finland, France, Germany, Greece, Hungary, Ireland, Italy, Latvia, Lithuania, Luxembourg, Malta, the Netherlands, Poland, Portugal, Romania, Slovakia, Slovenia, Spain, Sweden and Switzerland.

mental welfare deserving of protection. However, green criminological perspectives emphasise that despite such progress, legal systems often remain constrained by anthropocentric paradigms that treat animals primarily as property or resources, limiting the scope of meaningful justice for non-human species (Nurse, 2013, 2017). Nurse (2013) argues that while wildlife laws and animal welfare statutes have incrementally advanced species justice, these reforms frequently fall short due to exemptions for commercial exploitation and the marginalisation of animal protection within mainstream criminal justice systems. The literature further documents the politicisation of wildlife law, where economic and industry interests can undermine or roll back protections, highlighting the ongoing struggle to embed species justice principles within legislative and policy frameworks (Environmental Justice Australia, 2023; Nurse, 2013). The challenges around the legal recognition of sentience are further developed in the following section.

5 Ongoing Challenges to the Legal Recognition of Sentience

The ACT's *Animal Welfare Act* simultaneously demonstrates both the possibilities and persistent limitations inherent in animal welfare reform (Pallotta, 2019). The legislative journey and ongoing implementation of this groundbreaking law illuminate the complex dynamics between advancing animal welfare and protecting entrenched industry interests that characterise similar debates globally.

During the ACT's parliamentary debates, these tensions became visible when the Liberal party's concerns reflected the broader industry anxiety that recognising animal sentience could 'potentially be interpreted as prohibitive against many acceptable, responsible, humane and common environmental, commercial and recreational activities' (Kotzmann, 2023). These proposed exemptions were ultimately rejected, with Labor arguing they would make the legislation 'more ambiguous' and 'could be seen to limit… sentience', and the Greens opposing them on the basis that they wanted to see change in these very practices. This parliamentary resistance to industry-friendly exemptions represents a rare instance where progressive animal welfare advocates successfully resisted legislative capture (Kotzmann, 2023).

The ACT experience, however, must be understood within the broader pattern of industry opposition that has consistently shaped animal welfare policy both nationally and internationally. The animal agriculture industry's substantial economic influence and well-established lobbying mechanisms have enabled it to exert considerable pressure on policy-makers across jurisdictions, often resulting in resistance to changes that could potentially impact operations or profitability. This influence operates through various mechanisms, from direct lobbying to the strategic use of codes of practice that effectively exempt agricultural practices from animal welfare protections.

Globally, this pattern of industry resistance repeats with striking consistency. Broader scholarship indicates that the animal agriculture sector has historically leveraged government subsidies and favourable regulatory frameworks while resisting reforms that would impose stricter welfare standards or recognise animal sentience in law (Browning, 2022; Lazarus et al., 2022). Similarly, in the EU, agri-food lobbies have played prominent roles in diluting or delaying policy initiatives aimed at improving animal welfare, with powerful industry actors launching campaigns that frame animal welfare measures as threats to food security and economic stability (Pantzar, 2022). The European Parliament's agriculture committee, heavily represented by farming interests, has frequently sided with industry positions, resulting in stalled or compromised reforms despite widespread public support for improved animal welfare (Greenpeace European Unit, 2018; Kikou, 2024). Even in the United Kingdom, where an independent animal sentience committee was established to ensure government departments consider animal sentience in policymaking, concerns have emerged about industry capture when leadership positions are filled by individuals with direct industry ties, potentially undermining the committee's ability to advocate for meaningful change (Browning, 2022; Compassion in World Farming, 2021).

Despite the failure of the amendments noted above, the ACT's progressive stance is undermined by regulatory exemptions that perpetuate what green criminologists identify as systemic speciesism. Codes of practice continue to allow routine procedures that would constitute cruelty if performed on companion animals—including debeaking of poultry, confinement of sows in farrowing crates, and intensive housing of livestock under standards deemed 'humane' regardless of actual welfare impacts (Kotzmann, 2023; White, 2007). These exemptions reflect structural biases in enforcement (Morton et al., 2021) and represent a form of legislative capture where economic interests override species justice, as green criminological critiques emphasise (Sollund, 2021; White, 2007).

This regulatory framework demonstrates how even progressive animal sentience recognition can coexist with practices that fundamentally contradict the principles of sentience and intrinsic value. The ACT's experience reveals that statutory recognition of animal sentience, while symbolically important, does not automatically translate into comprehensive protection when industry exemptions remain embedded in the regulatory architecture. The tension between advancing animal welfare and protecting industry interests creates compromises that dilute legislative intent, highlighting the complexity of reforming animal welfare laws in environments shaped by competing interests and entrenched practices. Scholars note how tensions between the Act's progressive language and exemptions for industries illustrate the incremental nature of legal personhood advancements in anthropocentric systems (Kotzmann et al., 2023; Zanghellini, 2025).

The significance of the ACT case extends beyond its borders, offering both inspiration and caution for other Australian jurisdictions considering reform. Although explicit industry exemptions were resisted during the legislative process, the persistence of regulatory carve-outs through codes of practice illustrates how industry influence endures through alternative channels. The ACT's recognition of animal sentience and rejection of statutory exemptions mark an important advance, yet the survival of regulatory exemptions underscores the difficulty of converting symbolic recognition into substantive protection. This dynamic reveals how industry opposition adapts to context, shifting from direct parliamentary lobbying to regulatory capture to safeguard economic interests. Nonetheless, the ACT model remains a landmark for animal law reform, providing valuable lessons for states and territories navigating the tensions between progressive legislative ambition and entrenched industry resistance.

6 Rethinking Traditional Legal Classifications

As the first Australian jurisdiction to explicitly recognise animal sentience, the ACT sets a precedent for other states and territories to follow. The amendments align with international trends and respond to growing public concern that existing animal welfare laws inadequately protect animals from pain and suffering. The following provides two examples of how amended ACT *Animal Welfare Act* challenges traditional framings in the *Crimes Act*:

1. Recognition of Intrinsic Value and Mental Welfare: Under the *Animal Welfare Act*, animals are recognised as sentient beings with intrinsic value, deserving of humane treatment and consideration of their mental welfare—not just protection from physical harm. In contrast, the *Crimes Act* has historically treated animals as property, focusing almost exclusively on harms that have direct financial or ownership implications. For example, prosecutable cruelty under the *Crimes Act* typically relates to property damage or theft, whereas the ACT *Animal Welfare Act* foregrounds subjective experiences (pain, pleasure) and emotional welfare, shifting focus to the lived experiences of animals as individuals (Kotzmann, 2020).
2. Expanded Enforcement Powers for Animal Welfare: The *Animal Welfare Act* introduces an escalating enforcement framework, including interim prohibition orders and powers to seize, retain, rehome, or destroy animals to prevent further cruelty. This broadens the scope of enforcement far beyond the limited criminal penalties or restitution available under the Crimes Act, which does not contemplate animals' interests outside their status as property. For instance, the *Animal Welfare Act* enables action on abandonment and failure to provide mental stimulation (like failing to exercise a dog), acknowledging psychological and emotional suffering—forms

of harm unrecognised by the strict property-based lens of the Crimes Act.

Internationally, the recognition of animal sentience is reshaping legal landscapes, representing a pivotal shift towards acknowledging animals as subjects of moral concern rather than mere property. Scholars like Robertson and Goldsworthy (2022) argue that embedding sentience necessitates a duty to provide opportunities for positive affective states under the Five Domains model (Kotzmann, 2020). Peer-reviewed analyses highlight the symbolic and practical implications of these laws. Kotzmann (2020) argues that explicit sentience recognition provides an 'objective baseline' for evaluating human responsibilities towards animals, reducing arbitrariness in welfare judgements. In addition to the legislative shifts noted above, parliamentary debates on animal welfare legislation across various countries increasingly reveal a commitment to modernising legal frameworks in response to scientific advancements and evolving ethical standards (Englefield et al., 2019; Morton & Whittaker, 2022).

This shift is significant because it compels lawmakers and courts to move beyond utilitarian frameworks that focus solely on preventing cruelty, towards more robust obligations that reflect animals' intrinsic value and subjective experiences. As a result, legal systems are increasingly pressured to reconsider the adequacy of existing protections and to expand the scope of rights and responsibilities owed to animals. However, these changes also provoke complex debates about the boundaries of legal personhood, the extension of standing to animals in court, and the reconciliation of animal interests with entrenched economic and social practices. While sentience recognition holds symbolic power in rejecting animals' property status, its transformative potential remains constrained without parallel reforms to legal personhood, standing rights, and enforcement mechanisms (Kotzmann, 2020, 2023). Without advancing legal personhood for animals, their interests remain subordinated to human property rights, limiting the law's ability to prioritise animal welfare in practice.

Similarly, without granting animals or their advocates standing rights, sentience provisions risk being unenforceable, as violations may go unchallenged in court. Effective enforcement mechanisms are equally essential; legislative recognition alone cannot prevent harm unless backed by clear duties, penalties, and oversight. These limitations highlight the risks of reliance on regulatory tools premised on anthropocentric hierarchies, as evidenced by inconsistent protections across species and the persistent prioritisation of human convenience in codes of practice (Kotzmann, 2023; Robertson & Goldsworthy, 2022). Thus, while embedding sentience in law marks a critical step towards species justice, holistic change demands dismantling the regulatory and cultural paradigms that normalise human dominance over non-human species (Kotzmann, 2020; Sollund, 2021). Ultimately, the success of sentient animal legislation depends on bridging the gap between moral recognition and legal reality, embedding animal welfare into the fabric of enforceable rights and responsibilities.

One potential avenue for advancing this agenda lies through the adoption of integrated frameworks such as 'One Welfare', which recognise that advancing animal welfare is inseparable from promoting human well-being and environmental sustainability. One Welfare is a holistic concept recognising the interdependence of animal welfare, human well-being, and environmental sustainability, emphasising that the health and welfare of animals are closely linked to those of humans and their shared environment. It encourages interdisciplinary collaboration among a variety of stakeholders, including veterinary professionals, environmental scientists, mental health professionals, and animal caretakers, to promote ethical, sustainable, and effective care that benefits both animals and people globally (Pinillos et al., 2016).

Conceptually, the One Welfare framework provides a crucial theoretical bridge that connects these practical welfare considerations to broader transformative legal concepts such as legal personhood for animals. While One Welfare establishes the empirical foundation demonstrating the interconnectedness of human, animal, and environmental well-being, the recognition of animal legal personhood would provide the jurisprudential framework necessary to protect these interconnected relationships. Legal personhood for animals represents more than symbolic recognition—it constitutes a fundamental shift from viewing animals as property to recognising them as subjects with inherent rights that cannot be arbitrarily dismissed in favour of human economic interests. This transformation aligns with the One Welfare approach by acknowledging that animals possess interests worthy of legal protection, not merely as instruments for human benefit, but as beings whose welfare is intrinsically connected to broader ecological and social systems.

Furthermore, the integration of One Welfare principles with legal personhood frameworks resonates deeply with analytical lenses of eco-justice and species justice, which challenge anthropocentric approaches to environmental and animal protection. Eco-justice acknowledges transgressions against humans (environmental justice), specific ecosystems (ecological justice), and animal and plant species (species justice), recognising that these forms of harm are interconnected and require holistic responses. Species justice specifically emphasises the rights of non-human animals within their ecological contexts and their intrinsic right to pursue their natural behaviours and life patterns without human interference. This perspective complements the One Welfare approach by providing the ethical foundation for understanding why the interconnections between animal welfare, human well-being, and environmental health matter from a justice standpoint—not merely as pragmatic considerations, but as moral imperatives that demand institutional and legal recognition.

Such an integrated approach might play an important role in addressing the complex interconnections between human and animal well-being, fostering holistic approaches to justice that transcend species boundaries while simultaneously advancing the legal and institutional frameworks necessary to protect these relationships. The convergence of One Welfare‘s practical interdisciplinary methodology with the transformative potential of legal personhood and the normative grounding provided by eco-justice and species justice frameworks offers a comprehensive pathway towards multispecies justice that recognises the inherent value and interconnectedness of all life within our shared planetary systems.

7 Looking Forward

The explicit recognition of animal sentience in the ACT's *Animal Welfare Legislation Amendment Act 2019* (ACT) represents more than legislative reform—it signals a paradigmatic shift towards multispecies justice that challenges anthropocentric legal frameworks and opens pathways for transformative ecological governance. This recognition, while groundbreaking, is most effectively understood as part of a broader constellation of reforms that together have the potential to reshape human-animal-environmental relationships through interconnected frameworks of One Welfare, legal personhood, multispecies justice, and green criminology.

The One Welfare approach provides the foundational architecture for this transformation by recognising the interconnections between animal welfare, human well-being and the environment. By adopting One Welfare principles, future reforms can move beyond siloed approaches to create integrated policies that address the complex relationships between species justice, environmental degradation, and social equity. This holistic perspective aligns with the UN Sustainable Development Goals and offers a practical framework for implementing sentience recognition across multiple policy domains.

The trajectory towards legal personhood for animals and environmental entities represents the next evolutionary step beyond sentience recognition. While the ACT's acknowledgement of animal sentience establishes their capacity for subjective experiences, legal personhood would fundamentally challenge the property-based frameworks that currently constrain animal protection. As jurisdictions globally increasingly recognize rivers, forests, and ecosystems as legal persons with inherent rights, Australia has the opportunity to extend this paradigm to include individual animals. Such recognition would provide animals increased rights, creating robust mechanisms for protecting their interests beyond current welfare protections.

Multispecies justice offers the philosophical and practical framework for reimagining justice itself, this approach extends this logic to encompass non-human voices in decision-making processes, challenging the violent regime that maintains social, environmental, and species-based injustices. Green criminology's eco-justice framework provides the analytical tools for examining how current legal systems perpetuate harm against animals and the environment while identifying pathways for systemic reform. This approach recognises that the treatment of domesticated animals amounts to harm and should come under the purview of criminology, expanding criminal justice concepts to encompass species-based violence. By applying green criminological analysis to animal sentience legislation, future reforms can address structural speciesism while developing more comprehensive responses to animal victimisation.

The convergence of these frameworks creates unprecedented opportunities for ecological jurisprudence that recognises 'the intrinsic value of ecosystems, natural entities and the biosphere as a whole' (Burdon, 2012). Australian environmental law reforms currently underway, including the development of *National Environmental Standards* and new regulatory frameworks, provide openings for embedding sentience recognition and multispecies considerations into environmental governance. The proposed reforms to replace the 1999 Commonwealth Act of the Australian Parliament, *Environment Protection and Biodiversity Conservation Act* (*1999*) (Cth), offer particular opportunities to integrate animal sentience with ecological protection through nature positive approaches that prioritise environmental outcomes over process-driven assessments.

Looking forward, the synthesis of these frameworks suggests several priority areas for transformative reform:

1. Institutional transformation through the establishment of independent animal welfare authorities and Ministerial recognition for animal welfare can reduce conflicts of interest while ensuring sentience recognition translates into meaningful protection.
2. Procedural rights for animals can ensure that their interests are represented in environmental and development decisions, moving beyond current welfare protections to active inclusion in governance processes.
3. Intergenerational justice considerations can extend sentience recognition to encompass future generations of both humans and animals, addressing the temporal dimensions of ecological justice.

The ACT's sentience recognition thus represents not an endpoint but a catalyst for broader transformation towards ecological governance. This transformation requires challenging arguments that underpin current legal systems whilst hopefully embracing Indigenous knowledge systems that recognise the agency and rights of more-than-human beings. The emerging inter-judicial dialogue around animal rights and environmental personhood suggests that legal transformation is already underway that could eventually reshape international environmental law.

Ultimately, the future of animal sentience legislation in Australia lies not in isolated welfare reforms but in its integration within broader ecological justice movements that recognise the interconnectedness of all life. By embedding sentience recognition within One Welfare frameworks, advancing towards legal personhood, embracing multispecies justice principles, and applying green criminological insights, Australia can lead global efforts towards truly transformative reform. This integrated approach offers the potential to move beyond anthropocentric legal paradigms towards governance systems that honour the intrinsic value and agency of all Earth beings, creating the foundation for a just and sustainable multispecies future.

References

Animal Legal Defense Fund. (2024). *2023 U.S. State Animal Protection Laws Ranking Report*. https://aldf.org/wp-content/uploads/2024/02/2023-U.S.-State-Animal-Protection-Laws-Ranking-Report-Animal-Legal-Defense-Fund.pdf

Animal Welfare Act 1992 (ACT).

Animal Welfare Act 1999 (NZ).

Animal Welfare Amendment Act (No 2) 2015 (NZ).

Animal Welfare Legislation Amendment Act 2019 (ACT).

Bahari, M. M., Awami, R. R., Bukula, C. L., Nicodemus, J. S., & Kivuyo, M. L. (2024). Tanzania Animal Welfare Policy and Legal Framework Review: An Analysis Zeroing on Donkey Slaughter and Hides Trade. *Tanzania Journal of Agricultural Sciences, 23*(2), 400–409.

Beaver, B. V. & Golab, G. C. (2023). *The Veterinarian's Guide to Animal Welfare*. Academic Press.
Bionet, N.Z. (n.d.). *Animal Welfare and Ethics of Pest Management in New Zealand.* https://www.bionet.nz/rules/animal-welfare-and-ethics-of-pest-management-in-new-zealand.
Birch, J., Burn, C., Schnell, A., Browning, H., & Crump, A. (2021). *Review of the Evidence of Sentience in Cephalopod Molluscs and Decapod Crustaceans.* London School of Economics.
Blokhuis, H. J., & Veissier, I. (2023). Animal Welfare Assessment. In A. Knight, C. Phillips, P. Sparks, *Routledge Handbook of Animal Welfare* (pp. 25–42). Routledge.
Brisman, A. (2014). Of Theory and Meaning in Green Criminology. *International Journal for Crime, Justice and Social Democracy, 3*(2), 21–34. https://doi.org/10.5204/ijcjsd.v3i2.173
Broom, D. M. (2014). *Sentience and Animal Welfare*. CABI.
Broom, D. M. (2022). Animal Welfare Concepts. In A. Knight, C. Phillips, & P. Sparks (Eds.), *Routledge Handbook of Animal Welfare* (pp. 12–21) Routledge.
Browning, H. & Veit, W. (2022). The Sentience Shift in Animal Research. *The New Bioethics, 28*(4), 299–314.
Burdon, P. D. (2012). A Theory of Earth Jurisprudence. *Australian Journal of Legal Philosophy, 37*, 28–47.
Canali, E., & Keeling, L. (2009). Welfare Quality® Project: From Scientific Research to on Farm Assessment of Animal Welfare. *Italian Journal of Animal Science, 8*(sup 2), 900–903. https://doi.org/10.4081/ijas.2009.s2.900
Celermajer, D., Schlosberg, D., Rickards, L., Stewart-Harawira, M., Thaler, M., Tschakert, P., Verlie, B., & Winter, C. (2021). Multispecies Justice: Theories, Challenges, and A Research Agenda for Environmental Politics. In G. Hayes, S. Jinnah, P. Kashwan, D. M. Konisky, S. Macgregor, J. M. Meyer, & Zito, A. R. (Eds.) *Trajectories in Environmental Politics* (pp. 116–137) Routledge. https://doi.org/10.4324/9781003213321
Compassion in World Farming. (2021). *Written Evidence Submitted by Compassion in World Farming* (AWB0011). UK Parliament Committees. https://committees.parliament.uk/writtenevidence/37431/pdf/
Contreras, C. (2016). Sentient Beings Protected by Law: Analysis of Recent Changes in Colombian Animal Welfare Legislation. *Global Journal of Animal Law, 2*, 1–17. https://journal.fi/gjal/article/view/148716
Crimes Act 1900 (ACT).

Dentith, A. M., & Thompson, O. P. (2017). Teaching Adult Ecojustice Education. *New Directions for Adult and Continuing Education*, 2017(153), 65–75. https://doi.org/10.1002/ace.20222

Duffy, R., Hutchinson, A., Iordachescu, G. & Lappe-Osthege, T., (2025). A Harms-Based Political Ecology: Understanding Harms Through the Wildlife Trade. *Journal of Political Ecology 32*(1): Article 6226. https://doi.org/10.2458/jpe.6226

Englefield, B., Blackman, S. A., Starling, M., & McGreevy, P. D. (2019). A Review of Australian Animal Welfare Legislation, Regulation, Codes of Practice, and Policy, and Their Influence on Stakeholders Caring for Wildlife and the Animals for Whom They Care. *Animals, 9*(6), 335. https://doi.org/10.3390/ani9060335

Environment Protection and Biodiversity Conservation Act 1999 (Cth).

Environmental Justice Australia. (2023). *Concerns Over Environment Law Rollbacks.* https://envirojustice.org.au/legal-explainer-update-concerns-over-environment-law-rollbacks/

EUR-Lex. (n.d.). *Animal Welfare*. https://eur-lex.europa.eu/summary/glossary/animal_welfare.html

Evans, J. (2019). *Animal Sentience Recognised for the First Time Under New ACT Laws.* https://www.abc.net.au/news/2019-09-26/act-animal-sentience-recognised/11550946

Fear Free. (2025). *FAS Spectrum and Pain Algorithm.* https://fearfreepets.com/fas-spectrum/

Fischer-Tenhagen, C., Meier, J., & Pohl, A. (2022). Do Not Look at Me like That: Is the Facial Expression Score Reliable and Accurate to Evaluate Pain in Large Domestic Animals? A Systematic Review. *Frontiers in Veterinary Science, 9*, 1002681–1002681.

Franks, P., & Booker, F. (2018). *Governance Assessment for Protected and Conserved Areas (GAPA): Early Experience of a Multi-Stakeholder Methodology for Enhancing Equity and Effectiveness*. IIED. https://www.iied.org/17632iied

Fraser, D. (2008). The Role of the Veterinarian in Animal Welfare. Animal welfare: Too Much or too Little? Abstracts of the 21st Symposium of the Nordic Committee for Veterinary Scientific Cooperation (NKVet). Vaerløse, Denmark. September 24–25, 2007. *Acta Veterinaria Scandinavica, 50*(Suppl 1), S1–S12. https://doi.org/10.1186/1751-0147-50-s1-s1.

Gacek, J., & Jochelson, R. (2020). Animals as Something More than Mere Property: Interweaving Green Criminology and Law. *Social Sciences, 9*(7), 122–128. https://doi.org/10.3390/socsci9070122

Gibson, M. C. (2022). *Action-Oriented Studies in Green Criminology and a Harmful Trade in Pet Wildlife in Trinidad and Tobago* [Doctoral dissertation]. Michigan State University.

Greenpeace European Unit. (2018, May 23). Investigation: Majority of EU Parliament Agriculture Committee Linked to Industry. Greenpeace. https://www.greenpeace.org/eu-unit/issues/nature-food/1161/investigation-majority-of-eu-parliament-agriculture-committee-linked-to-industry/

Hampton, J. O., Hemsworth, L. M., Hemsworth, P. H., Hyndman, T. H., & Sandøe, P. (2023). Rethinking the Utility of the Five Domains Model. *Animal Welfare*, 32–42. https://doi.org/10.1017/awf.2023.84

Johnson, D. S. (2017). The Status of Green Criminology in Victimology Research. *McNair Scholars Research Journal, 10*(8).

Kikou, O. (2024). *New Report Augurs Further Delay of the Promised EU Animal Welfare Revolution.* EU Reporter. https://www.eureporter.co/environment/animal-welfare/2024/09/05/new-report-augurs-further-delay-of-the-promised-eu-animal-welfare-revolution/

Kotzmann, J. (2020). Recognising the Sentience of Animals in Law: A Justification and Framework for Australian States and Territories. *Sydney Law Review, 42*(3), 281–310.

Kotzmann, J. (2023). A Typology of Australian Animal Sentience Recognition Provisions—Enacted and Proposed. *Federal Law Review, 51*(2), 157–181. https://doi.org/10.1177/0067205X231166707

Kotzmann, J., & Stonebridge, M. (2021). There is Value in Stating the Obvious: The Case for Explicit Recognition of Animal Sentience in Legislation. *Cornell Law School Community*, 1–32. https://community.lawschool.cornell.edu/wp-content/uploads/2021/11/Kotzmann-Stonebridge-final.pdf

Kotzmann, J., Stonebridge, M., & Christiaenssen, P. (2023). Evolving Conceptions of Legal Personhood: What Might Recent Legal Developments Herald for Non-Human Animals in Australia? *UNSW Law Journal, 46*(2), 347–368.

Lazarus, O., McDermid, S.S., & Jacquet, J. (2021). The Climate Responsibilities of Industrial Meat and Dairy Producers. *Climatic, 165*(30), 1–23, https://doi.org/10.1007/s10584-021-03047-7

Lessard, M. (2023). *A Field Trip Into Québec Law: Exploring the Theoretical Ramifications of Sentience Recognition.* [unpublished manuscript] SSRN. https://papers.ssrn.com/sol3/papers.cfm?abstract_id=5092539

Martusewicz, R. A., Edmundson, J., & Lupinacci, J. (2015). *EcoJustice Education: Toward Diverse, Democratic, and Sustainable Communities* (2nd ed.). Routledge.

McEwan, A., & Turley, E. L. (2022). Green Criminology-Law Interdisciplinarity Towards Multispecies Justice: The Case of Wildlife Trafficking in Vietnam. *International Journal for Crime, Justice and Social Democracy, 11*(3), 92–106. https://doi.org/10.5204/ijcjsd.1804

Mellor, D. J. (2016). Updating Animal Welfare Thinking: Moving Beyond the 'Five Freedoms' Towards 'A Life Worth Living'. *Animals, 6*(3), 21–42. https://doi.org/10.3390/ani6030021

Mellor, D. J., & Beausoleil, N. J. (2015). Extending the 'Five Domains' Model for Animal Welfare Assessment to Incorporate Positive Welfare States. *Animal Welfare, 24*(3), 241–253. https://doi.org/10.7120/09627286.24.3.241

Mellor, D. J., & Reid, C. S. W. (1994). Concepts of Animal Well-Being and Predicting the Impact of Procedures on Experimental Animals. *Improving the Well-Being of Animals in the Research Environment*, 3–18.

Morton, R., Hebart, M. L., Ankeny, R. A., & Whittaker, A. L. (2021). Assessing the Uniformity in Australian Animal Protection Law: A Statutory Comparison. *Animals, 11*(1), 35. https://doi.org/10.3390/ani11010035

Morton, R., & Whittaker, A. L. (2022). Understanding Subordinate Animal Welfare Legislation in Australia: Assembling the Regulations and Codes of Practice. *Animals, 12*(18), 2437.

New Zealand Parliament. (2015). *Animal Welfare Amendment Act (No. 2) 2015*.

Nurse, A. (2013). Species Justice: The Future Protection of Wildlife and the Reform of Wildlife Laws. *The Green Criminology Monthly, 6*, 1–11. https://www.greencriminology.org/monthly/feb2013/The%20Green%20Criminology%20Monthly%20-%20February%202013%20-%20Species%20Justice%20-%20The%20Future%20Protection%20of%20Wildlife%20and%20the%20Reform%20of%20Wildlife%20Laws%20by%20Angus%20Nurse.pdf

Nurse, A. (2017). Green Criminology: Shining a Critical Lens on Environmental Harm. *Palgrave Communications, 3*(10), 1–4. https://doi.org/10.1057/s41599-017-0007-2

Pallotta, N. (2019). *Australian Capital Territory Enacts New Law recognising Animal Sentience*. Animal Legal Defense Fund. https://aldf.org/article/australian-capital-territory-enacts-new-law-recognizing-animal-sentience/

Pantzar, Y. (2022). *Animal Welfare: The EU Parliament Must Go Beyond the Interests of Farm Lobbies*. Slow Food. https://www.slowfood.com/blog-and-news/animal-welfare-the-eu-parliament-must-go-beyond-the-interests-of-farm-lobbies/

Pierrette, M., Honeckman, L., Jokela, F., Dunham, A., & Overall, K. L. (2023). Using Standardized Scales to Assess Fear at Veterinary Visits: Intra– and Inter-Rater Reliability. *Journal of Veterinary Behavior, 62*, 12–17. https://doi.org/10.1016/j.jveb.2023.02.004

Pinillos, R. G., Appleby, M. C., Manteca, X., Scott-Park, F., Smith, C., & Velarde, A. (2016). One Welfare—A Platform for Improving Human and Animal Welfare. *Veterinary Record, 179*(16), 412–413. https://doi.org/10.1136/vr.i5470

Jean-Loup Rault, Melissa Bateson, Alain Boissy, Björn Forkman, Bjørn Grinde, Lorenz Gygax, Jes Lynning Harfeld, Sara Hintze, Linda J. Keeling, Lubor Kostal, Alistair B. Lawrence, Michael T. Mendl, Mara Miele, Ruth C. Newberry, Peter Sandøe, Marek Špinka, Alex H. Taylor, Laura E. Webb, Laura Whalin, Margit Bak Jensen (2025). A Consensus on the Definition of Positive Animal Welfare. *Biology Letters, 21*(1). https://doi.org/10.1098/rsbl.2024.0382

Roberts, L. (2019, September 26). ACT Legislation Now Recognises Pets as “Sentient Beings” – ACT Rescue and Foster Inc. (ARF). Foster Dogs. https://fosterdogs.org/pets-recognised-as-sentient-as-government-adds-more-bite-to-animal-welfare-laws/

Robertson, I., & Goldsworthy, D. (2022). Recognising and Defining Animal Sentience in Legislation: A Framework for Importing Positive Animal Welfare Through the Five Domains Model. *Monash University Law Review, 48*, 244–269. https://doi.org/10.26180/21387387

Royal Society for the Protection of Animals Australia (RSPCA). (2019). Animal Welfare Legislation Amendment Bill 2019 RSPCA Australia Comments. https://kb.rspca.org.au/wp-content/uploads/2019/10/2019-02-08-RSPCA-comments-on-ACT-Animal-Welfare-Legislation-Amendment-Bill-2019.pdf

Schlosberg, D., Rickards, L., Pearse, R., Della Bosca, H., & Moraes, O. (2025). Critical Environmental Justice in Contemporary Scholarship and Movements: Consensus and Plurality of the Discourse. Environmental Politics, 34(3), 399–420. https://doi.org/10.1080/09644016.2024.2362573

Sollund, R. (2021). Nonspeciesist Criminology, Wildlife Trade, and Animal Victimization. In H. N. Pontell (Ed.), *Oxford Research Encyclopedia of Criminology*. Oxford Academic.

Taylor, P. S., Schrobback, P., Verdon, M., & Lee, C. (2023). An Effective Environmental Enrichment Framework for the Continual Improvement of Production Animal Welfare. *Animal Welfare, 32*, e14. https://doi.org/10.1017/awf.2023.5

Uhrig, P. W., & Robinson, N. G. (2006). *Canine Acute Pain Scale*. https://vetmedbiosci.colostate.edu/vth/wp-content/uploads/sites/7/2020/12/canine-pain-scale.pdf

White, R. (2013). *Environmental Harm: An Eco-Justice Perspective*. Policy Press.

White, R. (2016). Eco-Justice and the Moral Fissures of Green Criminology. In J. Jacobs and J. Jackson, (Eds.) *The Routledge Handbook of Criminal Justice Ethics* (pp. 76–94). Routledge.

White, R. (2018). Green Victimology and Non-Human Victims. *International Review of Victimology, 24(2)*, 239–255. https://doi.org/10.1177/0269758017745615

White, S. (2007). Regulation of Animal Welfare in Australia and the Emergent Commonwealth: Entrenching the Traditional Approach of the States and Territories or Laying the Ground for Reform? *Federal Law Review, 35*(3), 347–374. https://doi.org/10.22145/flr.35.3.1

Wilkins, C. (2019). *The Australian Capital Territory Recognises Animal Sentience in Law*. Horses and People. https://horsesandpeople.com.au/the-australian-capital-territory-recognises-animal-sentience-in-law/

Willis, R. S., Dunston-Clarke, E., Keating, L. R., Fleming, P. A., & Collins, T. (2021). Australian Livestock Export Industry Workers' Attitudes Toward Animal Welfare. *Animals, 11*(5), 1411. https://doi.org/10.3390/ani11051411

World Organisation for Animal Health (WOAH) OIE. (2025). *Animal Welfare*. https://www.woah.org/en/what-we-do/animal-health-and-welfare/animal-welfare/

Zanghellini, A. (2025). Affirming Animal Rights, Anthropocentrically. *Jurisprudence*, Advance online publication. https://www.tandfonline.com/doi/full/10.1080/20403313.2025.253360

8

Cities of Fires and Floods: A Spatial Justice Approach

Kajsa Lundberg

1 Introduction

Growing up in Sweden, I had limited experiences with environmental disasters before arriving in Naarm (named Melbourne by settlers on these lands) in 2017. Hence, while I was acutely aware of the growing impacts of climate change around the world, environmental disasters, their risks, and disaster preparedness never formed part of my life. A few years later, I had experienced smoke from one of Australia's largest bushfire events in settler history and lost my home to a flash flooding event. Parallelly, the implications of how and where we build on these fire- and flood-prone lands had become tremendously clear to me through my research. In drawing from these lessons, in this chapter I consider how green criminological concerns are at play in and around the flammable and floodable city.

K. Lundberg (✉)
Centre for Urban Research, School of Global, Urban and Social Studies, RMIT University, Melbourne, VIC, Australia
e-mail: kajsa.olivia.lundberg@rmit.edu.au

R. Killean et al. (eds.), *Conceptualising Justice*, Palgrave Studies in Green Criminology, https://doi.org/10.1007/978-3-032-19791-7_8

As a result of our changing climate, Australian cities, like many cities around the world, are facing increased challenges associated with extreme weather/disaster events such as bushfires and floods. Bushfire seasons are longer and more severe than before, rainfall varies significantly, and heavy precipitation events are becoming more intense (The Bureau of Meteorology and CSIRO, 2024). In this chapter, I have chosen to employ the term 'environmental disasters' to describe extensive fires and floods coming into contact with human infrastructure. I prefer this term over 'natural disasters', favoured by many governments, which I suggest overlooks the impacts of anthropogenic climate change. Moreover, to an extent, these lands have always been prone to fires, and water has always flowed here. Oftentimes, environmental disasters are the result of shortsighted planning that disregards the natural processes of other entities and life forms. Our cities appear largely unprepared for such extreme weather events and urban development patterns continue to place more-than-human lives at risk of associated harm. Despite the focus on human harm in this chapter, the concept of more-than-human is used here to signify the inseparability between human and non-human harms and well-being (see Whatmore, 2004). Other times, I refer specifically to human or non-human species, using those terms.

Spatial justice, building on work on the right to the city and the social production of space, brings together justice/injustice concerns with spatial and geographical elements such as access to safe and liveable spaces, environments, mobility, and exclusion/inclusion (Soja, 2009, 2010). As such, it concerns the social, political, and economic inequalities that structure our socio-spatial lives. The concept applies to work seeking more equitable cities, including issues around affordable housing, public transport systems, and urban population patterns. In this chapter, I employ spatial justice, more commonly associated with urban studies and geography, to draw attention to urban risk concerns associated with green criminological work on more-than-human harm and extreme weather events. This involves considerations of the distribution of and responses to environmental disaster risks and prevention for people living in different areas of the city.

More specifically, drawing from insights, policies, and reports associated with various recent fire and flood events in Australia and California,

USA, I sketch out the spatial injustices associated with urban political and planning responses as they relate to flood and bushfire risk. Reflecting on the findings from three public inquiries into Australian environmental disasters following the 2009 Black Saturday fires and the Black Summer fires (two recent severe fire events in Australia) as well as the 2022 Victorian floods (a severe flash-flooding event affecting various towns in the state of Victoria, one of Australia's eight states and territories), I consider the implications for urban spatial justice. As urban populations continue to grow, climate change accelerates, and extreme weather events become more frequent and severe, these challenges will intensify. In the next section, I start by detailing spatial justice approaches to date, including their application in (green) criminology, before turning to my analysis of the case studies.

2 On Spatial Justice

Early planning was associated with a belief in its deterministic potential, such as the ways efficiency, beauty, and 'good' could be designed into the city. However, the unforeseen and negative impacts of planning on lower socio-economic groups have led to scholars questioning what a 'good city' should look like (Fainstein, 2009). For example, Gray and Porter (2015) have demonstrated the disruptive impacts of urban regeneration during large sporting events on working-class communities. As a result, Fainstein (2009) proposed the concept of the 'just city'—informed by classical theories of justice—as the goal of planning. Urban planning, according to this perspective, represents a political and intentional act against unequal spatial dynamics.

Parallelly in geography, the concept of 'spatial justice' emerged, emphasising geography and space in the analysis of justice (Soja, 2009, 2010). In this view, central (in)justice elements (such as social, economic, and racial justice concerns) are embedded within a person's access to spaces and their affordances. It is 'a way of looking at justice from a critical spatial perspective' (Soja, 2009: 2–3). That is not to say that spatial justice should replace other forms of justice. Rather, according to Soja, it should complement these via its emphasis on the spaces that we live our lives in. Importantly, elements such as class, race, and gender are central determinants of what spaces are made available to different people (Soja, 2009, 2010).

In criminology, concerns with urban organisation, community resource distribution, and disorderly and criminal behaviour date back to the early twentieth century Chicago school and social ecology theories examining the environmental factors that contribute to crime (for example, Burgess, 1925/2020). Criminologists and criminal justice scholars today continue to consider spatial matters in the streets. For example, Fiddler (2018) and Kindynis (2017) have explored how violent acts linger (hauntingly) in space. Popovski and Young (2023) and Di Ronco (2023) have studied the policing and governance of protests in urban streets. Importantly, Di Ronco (2024) has also extended scholarship on urban governance to the control and management of non-human animals in the city. The exclusion of 'wild' animals from urban spaces, in line with urban governance more generally, Di Ronco (2024: 361) argues, tends to be in the interest of middle-class sensibilities, aesthetics, and consumption. Finally, Lundberg et al. (2024: 8) 'more-than-spatial' justice approach, drawing from Aboriginal peoples and the concept of 'Country', highlights the importance of land, air, water, spirituality, and more-than-human lives in spatial justice debates (see Nicholson & Jones, 2020). Drawing from this developing literature, in this chapter, I highlight our encounters with space as the source of various harms and (in)justices.

Green criminological engagements with spaces and justice tend to centre non-human victims of harm. For example, green criminologists have highlighted concerns around 'ecocide', the intentional or negligent killing of nature (e.g. White, Ch. 2 in this volume). Another dominant justice approach is environmental justice (EJ) that highlights different access to healthy, toxic, and polluted environments and their associations with race and poverty (Lynch & Long, 2022). Relatedly, the 'climate justice' approach highlights the unequal consequences of climate change (Brisman & South, 2018). Green criminologists have also started exploring spaces beyond Earth. Noriyoshi (2019), for example, has drawn attention to the market-driven mechanism and private endeavours that shape space exploration and capitalism, mirroring colonisation and occupation on Earth.

As such, green criminological justice concerns have contributed important insights related to inequalities, access, and harm to environments and more-than-human species. This includes a critique of speciesism alongside racism and the oppression of Indigenous peoples (South, 2014). This also involves political ecological critiques of late-stage capitalism's destruction of nature, its resources, and non-human life forms in favour of corporate interests and capital accumulation (Lynch & Long, 2022: 258). Such critiques also extend to 'the nature of private property rights versus the idea of the environment as a shared heritage to be held in common' (South, 2014: 8). These ideas and perspectives complement spatial justice debates by highlighting the socio-economic and political dynamics influencing spatial production.

To summarise, many of the concepts, topics, and concerns of green criminology develop and extend ideas prevalent in spatial justice debates. However, often the emphasis on spaces as dynamic entanglements between geographical space, affects, meanings, and power dynamics has been lost in criminological debates (Hayward, 2012), partially as a response to rational choice perspectives. Addressing this gap, influenced by spatial theorists (for example, Lefebvre, 1991), in this chapter, I approach space as socially produced through political influences, practices, creative expressions, and struggles. These spaces must be understood based on our phenomenological and sensory encounters with place (Lundberg, 2022a). According to this view, fire and flood events should be conceptualised as 'taking place in unstable and mobile spaces made up of complex arrangements of culture, meaning, politics, bodies and movements' (Lundberg, 2022b: 4).

A focus on velocity, time, and rhythm supports a more complex understanding of the relationships between spaces and our interpretations of fires and water flows in our everyday lives. For example, the speed and patterns of fire events are central in our interpretations of them as threats or as benign and valued climatic patterns (Lundberg, 2022b). Similarly, attention to deep time, describing the long history of humanity that preceded modern historical accounts, can significantly alter our reading of these events (Rademaker, 2021). A spatial justice approach drawing from such insights allows me to explore the past, present, and future planning and governance decisions that construct urban spaces as risky, and the ways that such risks are distributed spatially. By adopting a green criminological approach, I connect these issues with the topics of climate change and more-than-human co-existence. Next, I apply this approach to examples of fires and floods in Australia and California.

3 The Black Summer Fire: A Planning Failure

Both the scale and the reach of the 2019/2020 Australian Black Summer fires were extraordinary (Boer et al., 2020). Fires simultaneously burned in large parts of the country and in unusual areas of land such as rainforests (Nolan et al., 2020). Fires also appeared around large cities such as Melbourne, Canberra, and Sydney, covering these cities with blankets of toxic air (Morton, 2019). At least 33 people lost their lives in these fires, 2500 homes were destroyed (Huf & Mclean, 2020), and approximately 3 billion non-human animals were affected (WWF, 2020). These fires illustrate a significant and enduring shift in the patterns of Australian fire regimes, marked by increased intensity and duration (Nolan et al., 2020).

At the time of these fires, I was in the first year of my PhD, writing about the more-than-human harm implications of urban development and growth. That summer's fires, reported on around the world, spiked my interest in the relationships between bushfires and the ways we both design and live in cities. Tensions between urban growth and bushfires include building in high-risk areas and expanding electricity networks, which can ignite fires if they contact flammable materials like bush or fallen trees.

Concerns following the Black Summer fires included planning in the context of this fire-prone landscape and the impacts of climate change. The Royal Commission into National Natural Disaster Arrangements (Commonwealth of Australia, 2020) was established following the 2019/

2020 bushfires to examine 'natural disaster' risk, preparedness, and coordination. According to the inquiry findings, three primary factors can reduce the risk of environmental disasters: (1) a natural hazard involving natural processes, (2) the presence of people and other elements of value (3), and the vulnerability of individuals, communities, and their assets such as social, economic, or environmental factors (Commonwealth of Australia, 2020: 65–66). These are closely tied to planning and, in my reading of them, to spatial justice. Significant questions are: who are the people and communities that are living in bushfire-risky areas? And what are the planning processes, governance practices, and financial or social pressures that lead to people living in such fire-prone areas?

The public inquiry highlighted land-use planning as a mechanism for governments to manage natural disaster risk. This includes the importance of avoiding placing infrastructure and people in disaster-prone areas (Commonwealth of Australia, 2020: 70). This is particularly true for vulnerable human populations. One such vulnerability is the lack of understanding of fire risks and how to prepare for a potential blaze. The ways that land is used and kept, such as burning or slashing flammable areas, can also help reduce bushfire risk (Commonwealth of Australia, 2020: 366). Indigenous land management traditions are central to such practices. Indigenous Australians have a long history of caring for Country (Bawaka Country et al., 2015). One important strategy is the use of controlled fires to promote growth and reduce the excessive buildup of fuel in the landscape (Rose, 1996: 64). Clearly, the ability to burn areas of land to reduce the buildup of fuel in the landscape is prevented by the introduction of human infrastructure. Since the early twentieth century, urban settlements have led to a reduced inclination to burn in favour of fire suppression. The increased intensity, frequency, and spread of fires in recent years have been attributed to such fire suppression strategies and climate change (Gillson et al., 2019).

Land-use planning regimes regulate the locations of homes, businesses, and infrastructures, in part, to reduce disaster risk through strategic planning. Such planning is a responsibility shared between states, territories, and local governments. The zoning of land determines its allowed uses, such as housing and its density. Zones interact with overlays that set the standards for different types of development (Commonwealth of Australia, 2020: 400). Importantly, the ways we build today impact natural hazard risk decades, if not centuries, in the future. Extensive urban sprawl, for example, which comes to encroach on high-bushfire risk areas, tends to be associated with increased fire risk (Pyne, 2001/2020; Stanley & March, 2019). When people purchase a home, they inherit the riskiness of that space. As pointed out in the inquiry:

> Currently, all states permit homes to be built in bushfire and flood prone areas, and the degree to which planning or building standards act to mitigate risk varies across jurisdictions… there is still clear evidence of recent planning decisions placing communities at a known and obvious risk of disaster. (Commonwealth of Australia, 2020: 410)

Governments should protect more-than-human populations by preventing and mitigating known risks. Due to the pressures to provide affordable housing and meet developers' demands, building in high-risk areas may occur. The greater affordability of Melbourne's outer suburbs, similar to many global cities, attracts a higher proportion of low-income residents and immigrants. Although the rezoning of land for residential purposes might help provide people with affordable housing, in the context of a flammable landscape like that around Melbourne, this might be at the cost of the safety of such spaces. As demonstrated here, disaster risk tends to be unevenly distributed spatially across the city. In the next section, I consider the role of the electricity distribution network and powerful corporations in constructing fire risk in urban spaces.

4 Fire and the Electricity Distribution Network: Responsibility Relinquished

To understand the important relationship between the electricity network and fires, we must revisit another Victorian fire event that took place in 2009. These fires became known as the Black Saturday fires after the devastation that occurred on just one day. These fires started on the morning of Saturday February 7th, 2009, about 60 km north of

Naarm (Melbourne). This was the first of many blazes in Victoria on that day. The weather was extreme, following a long drought, with temperatures reaching 46°C (Hansen, 2018). Although these fires primarily affected rural and rural-urban areas, they also reached some suburban communities (Thornton, 2011).

In Australia, the risk of the electricity network sparking fires is notorious (Gill et al., 2013: 444–445). These fires often occur on extreme bushfire risk days when sparks can rapidly ignite large, fast-moving fires (Armistead et al., 2021). The Royal Commission Inquiry that followed Black Saturday established that at least five out of the eleven most significant fires were caused by failures in the electricity distribution network (due to, for example, electricity poles colliding or falling over due to wind) (Teague et al., 2010). A majority of all human deaths have also been attributed to electricity-caused fires (Thompson et al., 2013). Along with these discoveries, the Royal Commission established that large areas of the Victorian distribution network suffered from underinvestment and ageing distribution networks. Hence, the inquiry recommended that the network be upgraded (Teague et al., 2010: 151).

The Royal Commission's recommendations were later accepted by the Victorian government, agreeing to the replacement of the most vulnerable parts of the network, including SWER (single wire earth return) powerlines, commonly located in rural areas (Powerline Safety Taskforce, 2011). To support the implementation of the recommendations made by the Royal Commission, in 2010, the Powerline Bushfire Safety Taskforce was established. The taskforce recommended various measures to reduce the risk of bushfires caused from powerlines, including the instalment of protection technologies, the replacement of ageing SWER, and the use of underground or insulated overhead networks (Powerline Safety Taskforce, 2011: 9). Following the taskforce inquiry, the government established the Powerline Bushfire Safety Program worth $750 million to implement the taskforce's recommendations (D'Ambrosio, 2022). While this investment was on the lower end of what the Bushfire Safety Inquiry had recommended, the government sought to secure the most dangerous parts of the Victorian distribution network.

In this context, it is worthwhile to examine the political landscape surrounding the state of the electricity network in 2009. From the 1980s, neoliberalism proliferated in Australia and by the mid-1990s, the Victorian electricity supply was almost fully transferred to the hands of private companies (Nepal & Foster, 2015). This, it was argued at the time, would increase competition and reduce power prices while protecting quality. Nonetheless, evidence suggests that both quality and safety were sacrificed in the process (Armistead et al., 2021). With two major electricity distribution companies managing the infrastructure that delivers electricity to a majority of Victorian homes and businesses, it is questionable how that competition might play out in practice. In a report by Armistead et al. (2021), they established that the Victorian electricity distribution network is plagued by underinvestment and a lack of maintenance, with negative implications for network reliability, bushfire risk, and the safety of workers. Rather, the government has relinquished the responsibility for electricity to private companies in the pursuit of profit.

On a global scale, the recent January 2025 Southern Californian fires have once again highlighted the associations between urban sprawl, the rollout of electricity networks, and out-of-control fires. These fires were some of the worst in California's history. By late January, these were estimated to have caused the death of 29 people and the loss of thousands of homes and acres of burnt land, and to have significantly affected wildlife in the area (Karlamangla et al., 2025). In early January, fires started spreading around Los Angeles. The largest blaze was the Palisades fires, west of downtown Los Angeles, affecting the wealthy areas of Pacific Palisades, Malibu, and Santa Monica. The second largest was the Eaton fires that burned north of Pasadena in the San Gabriel Mountains (Karlamangla et al., 2025).

Significantly, in the aftermath of the California fires, victims and experts have directed blame at the electricity network for its role in causing these fires. Eyewitnesses claim to have seen flames around high-voltage transmission towers. As a result, victims sued Southern California Edison, the investor-owned public utility company servicing large parts of Southern California, for damages (Hals, 2025). Furthermore, based on data from Whisker Labs, a company claiming to monitor the safety of the electrical grid in the United States via innovative sensory technology, the power network in the areas of the Eaton, Palisades, and Hurst Fires, was under severe stress and various powerline faults coincided with the outbreak of the fires (Sacks, 2025). Southern California Edison denies responsibility for these fires (Hals, 2025).

The California fires also demonstrated that fires affect a diversity of people, including the more well-off. In this sense, spatial justice may differ from other approaches to justice, such as the environmental justice movement. Nonetheless, clear differences between people with more resources remain, including their abilities to protect themselves through insurance, to recover in the aftermath of disasters, and their abilities to relocate if needed. The harm and pain of losing your home and belongings should not be understated, however. Furthermore, this does not detract from the argument that governance decisions and the pursuit of profit contribute to the construction of the city and its spaces, and that such influences matter for urban justice. Moreover, private electricity companies pursuing profit over the safety of electricity users demonstrate the risks associated with financial pressures, profit incentives, and deregulation. As I move to consider flood risk in the next section, in addition to further emphasising financial pressures and planning choices, I highlight the importance of insurance companies and disaster risk understanding.

5 The Maribyrnong Floods: Commercial Interests in Planning

Almost three years after breathing the smoke from the Black Summer fires, the house I was renting in a middle-ring suburb of Melbourne was flooded in a flash flooding event. Floods around Victoria followed the wettest month on record, as rain fell over already wet ground. Sixty-three local government areas across the state flooded within a short period, causing the death of two men, and the destruction of hundreds of houses, businesses, livestock, crops, Aboriginal cultural heritage sites, and infrastructure (Legislative Council Environment and Planning Committee, 2024). Drone images in the media captured the devastation of entire towns, with the tops of houses and cars visible in the muddy waters. Here, special attention will be paid to Maribyrnong, a middle-ring suburb in Metropolitan Melbourne, where the Maribyrnong

River peaked at over four metres overflow, impacting the house that I was renting near the river.

I had been living in that house for over a year, and we had just signed to extend our lease as the river flooded in the early morning of October 14th, 2022. Before that, we had been oblivious to the fact that we were living in a known flood zone. As I watched the floodwaters creep up our inclined street, I realised for the first time that the newer houses had all been built on raised platforms. This flood, inundating 245 homes in the suburb of Maribyrnong (Bucci, 2022), was only the third worst since 1906. Therefore, many of the residents, in contrast to me, were aware of the risk. Still, almost 50 years had passed since the last major flood event in the area (Bucci, 2022). Our house, which had seen little renovation in the last decades, lacked any protections from the water.

Government investment into improving the resilience of existing buildings in high-risk areas (so-called legacy risks) includes raising homes, retrofitting, and buyback programs to support the relocation of people. Following the 2022 Victorian floods, a formal inquiry was set up to investigate the causes of the flood and its destructive forces as well as how to mitigate the risks of future flood events. The importance of resilience was raised by experts during the inquiry as a means of reducing the cost of future flooding events (Legislative Council Environment and Planning Committee, 2024). Resilience programs, however, are costly, demonstrating the importance of preventative planning as outlined in the last section. Therefore, the inquiry noted the significance of managing future flood risk through strategic land-use planning, including building controls in flood risk areas (Legislative Council Environment and Planning Committee, 2024: 133). Limiting new developments in areas prone to floods is central to minimising flood risk. Nonetheless, as Dominey-Howes (2022) writes in *The Conversation* and in the context of Sydney, the affordable housing crisis has entailed pressure to build in flood-risk areas. Various financial pressures encourage building in areas that jeopardise the safety of residents. Similar tensions exist in Melbourne and other major Australian cities that are struggling to provide housing to their growing populations.

Floods frequently occur on Victoria's flood plains, so many property owners in Maribyrnong were likely aware of the risk. This awareness could stem from personal or family experience with past floods or through discussions during property purchases and insurance agreements. Arguably the flood risk should have been made clear in my rental agreement, part of council communication, or perhaps I was responsible for staying informed. Regardless, certain lacks of knowledge can have negative implications, highlighting the importance of access to knowledge for disaster preparedness in achieving spatial justice.

The inquiry also established confusion amongst residents concerning the annual exceedance probability (AEP) that is used to communicate flood risk and, as a result, many underestimated the risk (Legislative Council Environment and Planning Committee, 2024). Stakeholders also raised the lack of adequate flood mapping and future modelling and noted the need to introduce modelling into the planning scheme. This is required to allow communities to understand the flood risks in the areas where they live or work. Moreover, currently, a person selling a property is not required to disclose flood risk. The inquiry recommended that the *Sale of Land Act 1962* (Vic) be altered to demand such disclosures by the vendors and that the government extend their communication with residents in flood zones. Furthermore, it established that disaster preparation plans need to include considerations of vulnerable communities, such as people with disabilities, and planning communication must be made available (and accessible) to all members of the community (Legislative Council Environment and Planning Committee, 2024). Knowledge around the risks associated with homes, including those inherited through the purchase of a property is central for people to make informed decisions around disaster risk and preparedness. Lack of such information, clearly, has implications for spatial justice.

Another point of contention brought up during the inquiry was the potential impact of the flood wall, built two decades earlier, around the Flemington Racecourse, a large horse racing venue nearby. Commercial interests had, in the building of the wall, it was argued during the inquiry, been allowed to trump the safety of the community. The racecourse's flood wall, critical voices claimed, contributed to the severity of the 2022 flood elsewhere. During the redevelopment of the Racecourse in the early 2000s, the flood wall was built. At that time, Melbourne Water had approved the wall, despite various objections concerning the impact of the wall on surrounding areas (Legislative Council Environment and Planning Committee, 2024). These objections were overturned by the Minister for Planning who highlighted the Racecourse's importance in strengthening Melbourne as a centre for finance, culture, tourism, commerce, and entertainment (Legislative Council Environment and Planning Committee, 2024: 151). Following the 2022 flooding of the Maribyrnong River, this decision was once more called into question by local residents, councils, and experts claiming that the flood wall contributed to the extensive damage from the 2022 floods (Legislative Council Environment and Planning Committee, 2024: 152). These alleged harms from the flood wall add to the abuse of animals in racing spaces which has more commonly been scrutinised by green criminologists. Although the inquiry determined the evidence inconclusive, it would be unsurprising if commercial interests were involved. The prioritisation of middle-class values and consumption in the design of cities and their governance is well established in criminology (Peršak & Di Ronco, 2021, 2024). It is important to interrogate the voices and influences that impact urban planning decisions, especially when these favour commerce and finance over the safety of residents.

Finally, I want to highlight insurance companies as crucial players in the aftermath of an environmental disaster. To fully grasp the role of insurance companies in urban governance and adaptation issues, a more comprehensive exploration than I can provide here is required. Nonetheless, I want to briefly highlight their role in relation to home insurance for those affected by events such as fires or floods. Even long after the flood, many residents of the Maribyrnong area continued to face homelessness and struggled to rebuild their lives. The inquiry heard from residents who had experienced delays from insurance companies, denied claims, low payouts, and increased premiums (Legislative Council Environment and Planning Committee, 2024). Delays or disputes of insurance claims, they found, exacerbated the challenges of those affected, leaving them to experience prolonged uncertainty and financial stress.

The inquiry also highlighted broader systemic issues, including gaps in insurance coverage, affordability concerns, and the inefficiency of processing claims after large-scale disasters (Legislative Council Environment and Planning Committee, 2024: 425). The important role of insurance companies includes the safeguarding of housing security amidst rising environmental uncertainties. Consequently, both insurance companies and the regulation of the insurance market are pivotal in advancing spatial justice.

This section has reinforced the importance of planning to consider environmental disaster risk and to seek alternative measures to satisfy urban housing needs. Various commercial pressures influence urban planning, affecting how communities are impacted in the aftermath of disaster events. Understanding flood risks and how structural elements, such as a flood wall, might exacerbate such hazards, is also central to improving spatial justice in urban areas.

6 Conclusion

In this chapter, I have employed spatial justice to uncover the unequal and calculated distribution of environmental disaster risk in urban areas, often in line with commercial interests and the pursuit of profit. I have provided examples of how urban development and planning interact with the risk of environmental disasters such as fires and floods. Moreover, I have detailed various elements that contribute to the oftentimes uneven distribution of environmental disaster risk across different sectors of the population, such as renters, homeowners, and people living in suburban and other high-risk areas. Moreover, I have shown how power dynamics, commercial interests, and consumption shape city planning and, consequently, how environmental disaster risk is integrated into urban spaces. Building on green criminological scholarship's expansion of concepts of spaces and more-than-human interactions—while bringing the production of space back into the analysis—I have demonstrated the importance of justice and environmental disaster risk in how we design our cities.

As urban populations continue to grow, climate change accelerates, and extreme weather events become more frequent and severe, the spatial justice concerns and challenges highlighted here will only intensify. We must also anticipate new and increased challenges associated with increased coastal flooding worldwide as sea levels rise (Alexander et al., 2012). This includes the complex dynamics associated with the impacts of environmental disasters on people of diverse socio-economic groups, as coastal floods spread beyond small island states to devastate the homes of prime ministers, CEOs, and judges along coastlines such as those surrounding Melbourne and Sydney. This underscores the importance of attending to space, temporality, and flexibility in our approaches to justice in urban spaces. As such, this chapter has demonstrated various spatial dynamics and their implications for justice both presently and in the future.

References

Alexander, K. S., Anthony, R., & Measham, T. G. (2012). Managed Retreat of Coastal Communities: Understanding Responses to Projected Sea Level Rise. *Journal of Environmental Planning and Management, 55*(4), 409–433. https://doi.org/10.1080/09640568.2011.604193

Armistead, A., Richardson, D., & Stanford, J. (2021). *Missing a Stitch in Time: The Consequences of Underinvestment in Proper Upkeep of Australia's Electricity Transmission and Distribution System.* The Australia Institute. https://australiainstitute.org.au/wp-content/uploads/2021/05/Missing-a-Stitch-in-Time-WEB-1.pdf

Bawaka Country, Wright, S., Suchet-Pearson, S., Lloyd, K., Burarrwanga, L., Ganambarr, R., Ganambarr-Stubbs, M., Ganambarr, B., & Maymuru, D. (2015). Working with and Learning from Country: Decentring Human Authority. *Cultural Geographies, 22*(2), 269–283. https://doi.org/10.1177/1474474014539248

Boer, M. M., Resco de Dios, V., & Bradstock, R. A. (2020). Unprecedented Burn Area of Australian Mega Forest Fires. *Nature Climate Change, 10*, 170–172. https://doi.org/10.1038/s41558-020-0716-1

Brisman, A., & South, N. (2018). Green Criminology and Environmental Crimes and Harms. *Sociology Compass, 13*(1), 1–12. https://doi.org/10.1111/soc4.12650

Bucci, N. (2022, October 21). *Did the Wall that Saved the Melbourne Cup Racetrack Contribute to the Flooding of 245 Homes?* The Guardian. https://www.theguardian.com/australia-news/2022/oct/21/did-the-wall-that-saved-the-melbourne-cup-racetrack-contribute-to-the-flooding-of-245-homes

Burgess, W. E. (1925/2020). The Growth of the City: An Introduction to a Research Project. In R. Park, W. E. Burgess, & Roderick D. McKenzie (Eds.), *The City Reader* (7th Ed.) (pp. 71–78). Routledge.

Commonwealth of Australia. (2020, October 28). *The Royal Commission into National Natural Disaster Arrangements Report*. https://www.royalcommission.gov.au/system/files/2020-12/Royal%20Commission%20into%20National%20Natural%20Disaster%20Arrangements%20-%20Report%20%20%5Baccessible%5D.pdf

D'Ambrosio, L. (2022, January 12). *Reducing Powerline Bushfire Risk to Communities.* Minister for Energy, the Environment and Climate Change. https://www.premier.vic.gov.au/sites/default/files/2022-01/220112%20-%20Reducing%20Powerline%20Bushfire%20Risk%20To%20Communities.pdf.

Di Ronco, A. (2023). *Policing Environmental Protest: Power and Resistance in Pandemic Times.* Policy Press.

Di Ronco, A. (2024). Human Control and 'Management' of Nonhuman Animals: New Research Directions for Green Criminology. *Crime, Media, Culture, 21*(3), 357–372. https://doi.org/10.1177/17416590241292070

Dominey-Howes, D. (2022). What's Causing Sydney's Monster Flood Crisis—And 3 Ways to Stop it From Happening Again. The Conversation, 3 July 2022, https://theconversation.com/whats-causing-sydneys-monster-flood-crisis-and-3-ways-to-stop-it-from-happening-again-186285

Fainstein, S. S. (2009). Planning and the Just City. In P. Marcuse, J. Connolly, J. Novy, I. Olivo, C. Potter, & J. Steil (Eds.), *Searching for the Just City: Debates in Urban Theory and Practice* (pp. 19–39). Taylor & Francis Group.

Fiddler, M. (2018). Ghosts of Other Stories: A Synthesis of Hauntology, Crime and Space. *Crime, Media, Culture, 15*(3), 463–477. https://doi.org/10.1177/1741659018788399

Gill, M. A., Stephens, S. L., & Cary, G. J. (2013). The Worldwide 'Wildfire' Problem. *Ecological Applications: A Publication of the Ecological Society of America, 23*(2), 438–454. https://doi.org/10.1890/10-2213.1

Gillson, L., Whitlock, C., & Humphrey, G. (2019). Resilience and Fire Management in the Anthropocene. *Ecology and Society, 24*(3). https://doi.org/10.5751/es-11022-240314

Gray, N., & Porter, L. (2015). By Any Means Necessary: Urban Regeneration and the 'State of Exception' in Glasgow's Commonwealth Games 2014. *Antipode, 47*(2), 380–400. https://doi.org/10.1111/anti.12114

Hals, T. (2025). *California Utility Faces Billions in Claims for Fire Damage Even If It Did Nothing Wrong*. Reuters, 15 Jan. 2025, sec. Energy. https://www.reuters.com/business/energy/california-utility-faces-billions-claims-fire-damage-even-if-it-did-nothing-2025-01-15/.

Hansen, C. (2018). Deep Time and Disaster. *Environmental Humanities, 10*(1), 226–240. https://doi.org/10.1215/22011919-4385543

Hayward, K. J. (2012). Five Spaces of Cultural Criminology. *The British Journal of Criminology, 52*(3), 441–462. https://doi.org/10.1093/bjc/azs008

Huf, B., & Mclean, H. (2020). *2019–20 Bushfires: Quick guide. Report for Department of Parliamentary Services, Parliament of Victoria*. Research Note no. 1, February. https://www.parliament.vic.gov.au/495ff4/globalassets/sections-shared/about/publications/research-papers/2019-20-bushfires_rn_2020_1.pdf

Karlamangla, S., Jiménez, J., Zhuang, Y., Selig, K., & Nostrant, R. (2025). *What We Know About the Wildfires in Southern California*. New York Times (Online). https://www.proquest.com/blogs-podcasts-websites/what-we-know-about-wildfires-southern-california/docview/3152417093/se-2?accountid=13552.

Kindynis, T. (2017). Excavating Ghosts: Urban Exploration as Graffiti Archaeology. *Crime, Media, Culture, 15*(1), 25–45. https://doi.org/10.1177/1741659017730435

Lefebvre, H. (1991). *The Production of Space* (Translated by Donald Nicholson-Smith). Blackwell.

Legislative Council Environment and Planning Committee. (2024). *The 2022 Flood Event in Victoria: Final Report*. Parliament of Victoria. https://www.parliament.vic.gov.au/495fa6/contentassets/a1a3e3c7d84c4a059b7f8ad12089af06/lcepc-60-02_2022-flood-event-in-vic_final-report.pdf

Lundberg, K. (2022a). Walking at the Edges of Green Criminology: The Edges of the City and the Extraordinary Consequences of Ordinary Harms. *Criminological Encounters, 5*(1), 18–34. https://doi.org/10.26395/CE22050103

Lundberg, K. (2022b). Moved by Fire: Green Criminology in Flux. *Crime, Media, Culture, 18*(1), 3–20. https://doi.org/10.1177/1741659020958450

Lundberg, K., Popovski, H., & Young, A. (2024). Justice in the Streets? Towards a Criminology of More-Than-Spatial Justice. *Criminology & Criminal Justice*, 17488958241257834. https://doi.org/10.1177/17488958241257834

Lynch, M. J., & Long, M. A. (2022). Green Criminology: Capitalism, Green Crime and Justice, and Environmental Destruction. *Annual Review of Criminology, 5*, 255–276. https://doi.org/10.1146/annurev-criminol-030920-114647

Morton, A. (2019, December 24). *Yes, Australia Has Always Had Bushfires: But 2019 is Like Nothing We've Seen Before.* The Guardian. https://www.theguardian.com/australia-news/2019/dec/25/factcheck-why-australias-monster-2019-bushfires-are-unprecedented.

Nepal, R., & Foster, J. (2015). Electricity Networks Privatization in Australia: An Overview of the Debate. *Economic Analysis and Policy, 48*(December), 12–24. https://doi.org/10.1016/j.eap.2015.10.001

Nicholson, M., & Jones, D. (2020). Wurundjeri-al Narrm-u (Wurundjeri's Melbourne): Aboriginal Living Heritage in Australia's Urban Landscapes. In K. Silva (Ed.), *The Routledge Handbook on Historic Urban Landscapes in the Asia-Pacific* (pp. 508–525). Routledge.

Nolan, R. H., Matthias, B. M., Collins, L., Resco de Dios, V., Clarke, H., Jenkins, M., Kenny, B., & Bradstock, R. A. (2020). Causes and Consequences of Eastern Australia's 2019–20 Season of Mega-Fires. *Global Change Biology, 26*(3), 1039–1041. https://doi.org/10.1111/gcb.14987

Noriyoshi, T. (2019). Astro-Green Criminology: A New Perspective Against Space Capitalism Outer Space Mining May Make the Same Mistakes in Space as we Have on Earth. *Toin University of Yokohama Bulletin, 40*, 7–16.

Peršak, N., & Di Ronco, A. (2021). *Harm and Disorder in the Urban Space: Social Control, Sense and Sensibility*. Routledge.

Peršak, N., & Di Ronco, A. (2024). Security Governance, Harms and Resistance: Recalibrating Security and Redefining Incivility Through Legitimacy Concerns. *Criminology & Criminal Justice*, 17488958241254445. https://doi.org/10.1177/17488958241254445

Popovski, H., & Young, A. (2023). Small Things in Everyday Places: Homelessness, Dissent and Affordances in Public Space. *The British Journal of Criminology, 63*(3), 727–747. https://doi.org/10.1093/bjc/azac053

Powerline Bushfire Safety Taskforce. (2011). *Powerline Bushfire Safety Taskforce: Final Report*. https://www.energysafe.vic.gov.au/sites/default/files/2022-12/PBST_final_report_30Sep2011.pdf

Pyne, S. J. (2001/2020). *Fire: A Brief History*. New South Publishing.

Rademaker, L. (2021). A history of deep time: Indigenous knowledges and deep pasts in settler-colonial presents. *History Australia, 18* (4), 658–675. https://doi.org/10.1080/14490854.2021.1988651

Rose, D. B. (1996). *Nourishing Terrains*. Australian Heritage Commission.

Sacks, B. (2025, January 10). *Did Power Lines Help Start the L.A. Fires? What We Know*. The Washington Post (Online). https://www.proquest.com/docview/3153964265/citation/6B1302D63BC441E1PQ/1

Soja, E. W. (2009). The City and Spatial Justice. *Justice Spatiale/Spatial Justice, 1*, 1–5.

Soja, E. W. (2010). *Seeking Spatial Justice*. University of Minnesota Press.

South, N. (2014). Green Criminology: Reflections, Connections, Horizons. *International Journal for Crime, Justice and Social Democracy, 3*(2), 5–20. https://doi.org/10.5204/ijcjsd.v3i2.172

Stanley, J., & March, A. (2019). *Black Saturday: Urban Sprawl and Climate Change Remain Key Dangers*. Pursuit. https://pursuit.unimelb.edu.au/articles/black-saturday-urban-sprawl-and-climate-change-remain-key-dangers

Teague, B., McLeod, R., & Pascoe, S. (2010). *2009 Victorian Bushfires Royal Commission Final Report: Summary*. State Government of Victoria. http://royalcommission.vic.gov.au/finaldocuments/summary/PF/VBRC_Summary_PF.pdf

The Bureau of Meteorology & CSIRO. (2024). *State of the Climate 2024*. http://www.bom.gov.au/state-of-the-climate/2024/documents/2024-state-of-the-climate.pdf

Thompson, G., Christodoulou, M., & Cronau, P. (2013, October 28). *Powerline Failures Scrutinised as Potential Cause of Bushfire Crisis*. ABC News. http://royalcommission.vic.gov.au/finaldocuments/summary/PF/VBRC_Summary_PF.pdf

Thornton, R. P. (2011). Short Communication on Research Response to the Black Saturday (7th February 2009). Victorian Bushfires, Australia. *Fire Technology, 47*(2), 295–301. https://doi.org/10.1007/s10694-010-0175-8

Whatmore, S. (2004). Humanism's Excess: Some Thoughts on the 'Post-Human/ist' Agenda. *Environment and Planning A, 36*(8), 1360–1363. https://doi.org/10.1068/a37127

WWF. (2020). *Australia's 2019–2020 Bushfires: The Wildfire Toll*. https://assets.wwf.org.au/image/upload/v1/website-media/resources/Animals_Impacted_Interim_Report_24072020_final?_a=ATO2Bcc0

9

The Neglected 99%: Towards Invertebrate Justice

Russil Durrant

1 Introduction

An important strand of green criminological scholarship focuses on our moral and legal responsibilities to non-human animals. In many respects, this represents the most radical rethinking of the subject matter of criminology since its inception as an academic discipline: if we reject the idea that criminology should focus only on criminal acts and only those which result in human victims, then it opens the possibility for exploring the harm we inflict on trillions of animals every year through both legal and illegal activities. Despite the growing interest in research on non-human animals within criminology (Wyatt, 2024), criminologists have

This work was supported by the Australian Research Council (grant number: DP230100630).

R. Durrant (✉)
Institute of Criminology, Victoria University of Wellington, Wellington, New Zealand
e-mail: russil.durrant@vuw.ac.nz

R. Killean et al. (eds.), *Conceptualising Justice*, Palgrave Studies in Green Criminology, https://doi.org/10.1007/978-3-032-19791-7_9

more or less ignored the most common group of animals on the planet—invertebrates. In a recent book (Durrant, 2024) I have argued for the extension of green criminological scholarship to include invertebrate life and, drawing from White's (2013) eco-justice framework, introduced the Invertebrate Justice Model to capture our direct and indirect justice obligations towards invertebrates. In short, this model recognises that we need to care more about invertebrates not only for their own sake, but also because their flourishing is crucial for the integrity of ecosystems that all species—including humans—rely upon. In this chapter, I reprise some of the core arguments of the book and apply them in the context of existing animal welfare law in Aotearoa New Zealand. In particular, I advance the idea that we should include insects under the *Animal Welfare Act 1999* (NZ) ('*Animal Welfare Act*'), bringing many trillions of individuals within the ambit of legal protection for the first time.

To make this case I draw from three important recent developments: the 'animal turn' in criminology (Wyatt, 2024), the growing importance of the concept of sentience in animal law (Blattner, 2019), and expanding research on invertebrate sentience in general, and insect sentience in particular (Gibbons et al., 2022; Mikhalevich & Powell, 2020). First, I note that invertebrates, despite making up most of animal life, have been neglected in a range of academic disciplines, including criminology. I then provide a precis of the Invertebrate Justice Model (Durrant, 2024), which outlines how we have both direct and indirect justice obligations towards invertebrates. In the second half of the chapter, I take one approach to advancing invertebrate justice: expanding animal welfare law in Aotearoa New Zealand to include insects. Although I will make the case that including insects is likely to be a positive development, ultimately a more radical reform of legal systems is needed to advance justice for non-human animals.

2 The Neglected 99%

'We are the 99%' was the rallying call of the Occupy movement in 2011. The aim, of course, was to draw attention to the sharp disparities in income and wealth between the richest 1 percentile and the rest of the population (Milkman et al., 2014). Reducing inequalities in wealth, both within and among societies, remains one of the most pressing issues of our time (Piketty, 2024). In invoking 'the neglected 99%' I want to draw attention, however, to a rather different group of individuals who—despite representing the majority—are nonetheless systematically under-represented: the invertebrates. The term invertebrate simply captures all of those animals that lack a vertebrate column, or backbone (Brusca & Brusca, 2003). Invertebrates include major groups of animals such as insects, crustaceans (e.g. crabs, lobsters and prawns), molluscs (e.g. octopuses, mussels, and clams), spiders and others. Although biologists do not have a precise understanding of the number of different species on our planet, more than 95% of the known animals are invertebrates (Eisenhauer & Hines, 2021). Given the ongoing discovery of new species and the fact that most of the animals that have yet to be formally described are invertebrates, the number is likely to be substantially higher (Stork, 2018). If we consider the number of *individual* animals alive at any given point in time, then almost certainly more than 99% are invertebrates.

Invertebrates, however, tend to be under-represented across a range of disciplines including biology, conservation studies, and the social and behavioural sciences (see Durrant, 2024). It is perhaps no surprise, then, that invertebrates are largely absent in criminology as well (Durrant, 2024). After all, mainstream criminology has paid relatively little attention to the harms that humans inflict on non-human animals in general (Beirne, 1999; Girgen, 2022; Taylor & Fitzgerald, 2018). However, the emergence of green criminology in the 1990s, and its commitment to non-speciesism (Sollund, 2021; White, 2018), have seen a growing group of scholars who have examined harms against animals in a range of contexts—from wildlife trafficking (Sollund, 2024) to illegal dog fighting (Nurse, 2021)—suggesting a possible 'animal turn' in criminology (Wyatt, 2024). The fact that criminologists have engaged in research on conventionally uncharismatic animals like sharks (Walters & Couper, 2023), and eels (Pons-Hernandez, 2024) provides optimism that criminologists will also begin to extend their concern to invertebrate species as well.

I think that there are two compelling reasons why criminologists in general, and green criminologists in particular, should pay more attention to invertebrates. First, if we accept the idea that criminology as a discipline should concern itself with harms, not just crimes, and if we reject an anthropocentric focus on harms against humans and recognise that non-human animals can also be victims (Arkow, 2021; White, 2018), then a case can be made that the harms that we inflict on invertebrates are one of the most serious moral problems of our time. This follows from the sheer scale of the harm that we cause invertebrates across a diverse range of contexts. We raise and slaughter huge numbers of invertebrates for food, feed and material products every year, exploit invertebrates in medical and research contexts, utilise the services of invertebrates to pollinate our crops, and kill very large numbers with pesticides and other chemical pollutants (Durrant, 2024; Rowe, 2020). Although estimates inevitably come with a large margin of error, something in the region of 10 quadrillion (i.e. 10 followed by 15 zeroes) invertebrate individuals may be deliberately killed by humans every year (Rowe, 2020). To put that number in perspective, many more invertebrates are killed *every year* than humans who will probably *ever* exist (e.g. Browning & Veit, 2022a). Critically, as I will elaborate on later in this chapter, emerging research suggests that many of the invertebrates killed by humans likely have the capacity for sentience. That is, they have the capacity to subjectively experience the world (to feel pleasure and pain) in ways that mean that their welfare needs to be taken into account (Browning & Veit, 2022b).

A second reason why I think we need to pay more attention to invertebrates in criminology is that their well-being and flourishing is not only important for their own sake, but also for the integrity and flourishing of the ecosystems that humans (and other animals) ultimately depend upon. Invertebrates provide critical 'ecosystem services' including pollination, disease control, nutrient recycling, soil formation, and water purification (Cardoso et al., 2020). The decline in invertebrate numbers is already having an impact on species that rely upon them for food (Goulson, 2021), and insect declines are affecting the pollination 'services' offered by bees, flies, and other species. One recent study, for example, estimated that through their impacts on human diets, pollinator declines are responsible for 427,000 excess human deaths annually (Smith et al., 2022). In sum, I argue that we need to pay more attention in criminology to *invertebrate justice* (Durrant, 2024).

3 The Invertebrate Justice Model

In the face of the current environmental polycrisis, there is a growing interest in considering justice for the more than human world, and a number of frameworks have been proposed to better understand our moral and legal obligations towards non-human species, and the environment. Within green criminology, Rob White's (2013, 2018) eco-justice framework has proven influential in shaping thinking about how we can better realise just outcomes for humans, non-human animals, and ecological systems. In brief, White (2013) proposes that there are three main domains of justice that arise from our interactions with the environment: *environmental justice* concerns justice for humans in relation to the environment; *species justice* focuses on justice for non-human organisms (especially animals); and justice for the environment, broadly construed, is the focus of *ecological justice*. Utilising this framework, I have proposed an Invertebrate Justice Model that focuses on these three justice domains as they relate to our interactions with invertebrate species (Durrant, 2024: Figure 1).

As illustrated in Fig. 1, the Invertebrate Justice Model has two main justice pathways. The first relates to our *direct* justice obligations towards individual invertebrates and falls within the domain of species justice. There is now an extensive literature in moral and political philosophy that has focussed on outlining what our moral obligations are towards non-human animals (e.g. Regan, 2004; Singer, 2023) and therefore how we should organise our social, legal, and political institutions to best realise those obligations (e.g. Cochrane, 2018; Donaldson & Kymlicka, 2011). Although there is a diversity of theoretical perspectives, one widespread idea is that the nature of our moral and justice obligations towards non-human animals depends in large part on their capacity for sentience. More specifically, many philosophers argue that sentience is both a necessary and sufficient condition for moral consideration (Perry, 2024). It should be noted, however, others argue that we also have moral and justice obligations towards non-sentient organisms as well (Sandler, 2018), while a multispecies justice perspective focuses more on how our justice concerns should be shaped by the *relationships* that emerge among the more than human world (Celermajer et al., 2020). While agreeing that we have some moral obligations towards both sentient and non-sentient invertebrates, I have argued that we have stronger moral, and especially justice, obligations to those invertebrates (and other animals) that we can say are probably sentient (Durrant, 2024). The case for prioritising sentience in our interactions with individual animals follows from the recognition that although we can cause 'harm' to both sentient and non-sentient animals, only sentient animals subjectively *experience* that harm and therefore have a 'welfare' that needs to be taken into consideration (Rowlands, 2025). As I shall outline in more detail below, an approach that prioritises direct justice concerns for sentient invertebrates is consistent with recent developments in animal law and offers scope to better recognise the legal status of invertebrates in ways that might be able to reduce harm (Blattner, 2019).

The second pathway in the Invertebrate Justice Model focuses on the indirect justice obligations that arise from our interactions with invertebrates. That is, it focuses on how our interactions with invertebrates can impact on the well-being and flourishing of humans (environmental justice), other species (species justice), and the environment (ecological justice). The concern here, therefore, is less with the harm directly experienced by individual invertebrates and more about how our interactions with invertebrates affect ecological systems in ways that can generate harm to humans, non-humans, and the environment. To illustrate, consider the current expansion of the krill fishing industry. Antarctic krill are small crustaceans that are a 'keystone species' in the Southern Ocean environment (Cavan et al., 2022). A case could be made that we have direct justice obligations towards individual krill. Although the capacity for sentience among krill is currently unknown, given their relative biological similarity to other crustaceans such as shrimps (who display some markers of sentience), it is entirely possible that they also have the

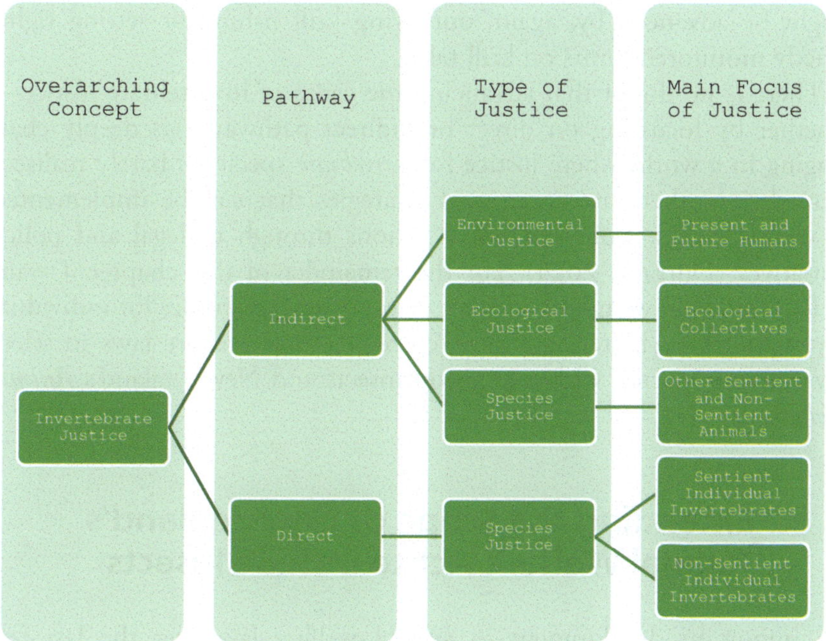

Fig. 1 The Invertebrate Justice Model (Durrant, 2024: 144)

capacity for sentience (Birch et al., 2021). If we take this possibility seriously, then it should mean adopting legally enforced measures to reduce the harm that we inflict on individual krill or, ideally, to outlaw krill fishing entirely. Our interactions with krill also lead indirectly to justice concerns for humans, other animals, and the environment. Because krill play an important role in sequestering carbon (Trebilco et al., 2020), their large-scale harvesting will contribute to global warming which, in turn, generates many significant harms for humans, especially those in the Global South (Cripps, 2022). A reduction in krill numbers will also have reverberating impacts on the Southern Ocean marine food chain as it will impact on the flourishing of all those individuals that rely on krill for food, including fishes, seals, penguins, and whales (Cavan et al., 2022). It follows that the flourishing of the marine environment more generally is adversely affected by large scale krill fishing. Here justice might be advanced by, again, outlawing krill fishing or setting tight, strictly monitored limits on krill takes.

There is no doubt that advancing the cause of invertebrate justice—whether by focussing on direct or indirect pathways—is deeply challenging in a world where justice for *vertebrate* species is barely realised. Nonetheless, there are a range of strategies that can be implemented at various scales, from individual actions through to legal and policy initiatives (Durrant, 2024). For the remainder of this chapter, I want to focus on how we might advance direct justice outcomes for individual invertebrates by better extending current animal welfare laws to select invertebrate species, with a focus on insects and New Zealand's *Animal Welfare Act*.

4 Extending the Scope of New Zealand's Animal Welfare Act to Include Insects

An important development in animal welfare law over the last few decades has been the increasing recognition of sentience as a key criterion for animal protection (Blattner, 2019; Kotzmann & Ferrere, 2024). As outlined above, there is substantial agreement among moral and political philosophers that sentience is a sufficient (and, perhaps, necessary) requirement for moral consideration (Blattner, 2024). If we need to consider the positive and negative hedonic states of non-human animals in our interactions with them, then it follows that legal provisions should be constructed in ways that promote practices that take these experiences into account (Blattner, 2019). The implicit or explicit legal recognition of sentience is a feature of animal welfare and animal cruelty laws in a number of jurisdictions, including many European countries, the United Kingdom, some Australian States, and Aotearoa New Zealand (Blattner, 2019, 2024; Kotzmann, 2023a). For example, changes to the New Zealand *Animal Welfare Act* were implemented in 2015 'to reform the law relating to the welfare of animals and the prevention of their ill-treatment; and, in particular,—(i) to *recognise that animals are sentient*' (preamble, emphasis added) (Ferrere, 2024).

As Blattner (2024) outlines, the legal recognition of animal sentience has a wide range of effects that work to shape the way that animal welfare laws are developed and implemented. First, and perhaps most importantly, by highlighting the importance of sentience as a criterion for legal protection, animal welfare laws—in principle at least—recognise the intrinsic value of animals and their specific needs and interests. This can help to circumvent the cognitive biases that often shape our interaction with other species in ways that lead to differential treatment of animals based on anthropocentric or human centred considerations like how attractive or useful they are, or how similar they are to humans (see Herzog, 2021; White, 2018). This, in turn, may form the basis for engaging with the public in ways that can advance a more robust recognition of fundamental animal rights that go beyond existing laws (Kotzmann, 2023b, 2024). One potentially important contribution of the inclusion of sentience in legal provisions is that it offers to broaden the scope of animals that might be included within the scope of the law.

Historically, animal welfare and animal cruelty laws have defined 'animals' in a way that has largely or completely ignored invertebrate species (Durrant, 2024). Currently, only a handful of jurisdictions extend animal welfare law to invertebrates and the range of invertebrates that are included is largely limited to cephalopod molluscs (octopuses, squid, and cuttlefish), and some crustacean species (especially those in the order of 'decapod crustaceans'). In the New Zealand *Animal Welfare Act* (s 2(1)(a)), for instance, the term 'animal' is used to refer to 'any live member of the animal kingdom that is (i) a mammal; or (ii) a bird; or (iii) a reptile; or (iv) an amphibian; or (v) a fish (bony or cartilaginous); or (vi) any octopus, squid, crab, lobster, or crayfish (including freshwater crayfish)'. Somewhat oddly, cuttlefish are omitted from this list, along with other crustacean species (e.g. shrimps) that are included in other jurisdictions (e.g. in the *Animal Welfare (Sentience) Act 2022* (UK)). However, perhaps of greatest note is the exclusion of all insect species from the definition of 'animal' and hence from any kind of legal protection that specifically relates to their welfare. Indeed, only Norway, to my knowledge, has included insects within its animal welfare laws and that is limited to one species: the honey bee (Blattner, 2019). The New

Zealand *Animal Welfare Act* does, however, explicitly state that the definition of 'animal' includes 'any other member of the animal kingdom which is declared from time to time by the Governor-General, by Order in Council... to be an animal for the purposes of this Act' (s 2(1)(a)(vii)). This provision, alongside the recognition that animals are sentient, offers an opportunity for Aotearoa New Zealand to include insects within its animal welfare legislation and hence advance justice for a wider range of invertebrates.

To make a robust case for the inclusion of some or all insects in New Zealand's *Animal Welfare Act* it will be important to argue that there is a sufficient likelihood that they are, indeed, sentient. The task of evaluating sentience in other species is a notoriously challenging one, because we only have direct knowledge of our own experiences. Nonetheless, it is reasonable to make claims about the consciousness or sentience of other beings by examining relevant similarities to animals that we can be more confident are sentient, especially humans (Andrews et al., 2025). In particular, we can look at a range of behavioural and neurobiological criteria that are likely to be markers of sentience and evaluate to what extent different animals meet these criteria (Birch et al., 2021; Gibbons et al., 2022). Relevant criteria include things like the possession of nociceptors (sensory receptors that respond to harmful stimuli), integrative brain regions, response to analgesia (e.g. painkillers), associative learning, flexible self-protection, and the capacity to engage in motivational trade-offs (see Birch et al., 2021 for details).

The idea of insect sentience has long been controversial, and one view has been that, although insects may engage in complex behaviour, they do not have the capacity to experience pain and pleasure (Durrant, 2024). That is, they are not sentient. A thorough review of the research on insect sentience is beyond the scope of this chapter. However, a recent analysis of research on eight potential criteria for sentience in six different insect orders concluded that evidence for these criteria was found in varying degrees across the insect orders studied with a high or very high degree of confidence (Gibbons et al., 2022). As can be seen in Table 1, this led the authors to conclude that there is strong to substantial evidence for sentience in many different insect species. Moreover, the evidence for sentience in insects is similar to that found for other invertebrates such as cephalopods and decapod crustaceans that are more widely included in animal welfare legislation. There does remain, however, a substantial degree of uncertainty regarding these findings and only a small number of insect species have been studied in detail. Complicating this picture is the fact that many insect species go through significant morphological changes from juveniles to adults—so called holometabolous species like butterflies, bees, flies, and ants, for instance, start out as larvae, develop as pupae and metamorphose into adults. This means that we need to evaluate their capacity for sentience in all relevant life stages (Gibbons et al., 2022). How, then, do we deal with uncertainty over the sentience of insects in making decisions about their legal status?

Many have argued that we need to take a *precautionary approach* (e.g. Birch, 2017, 2024; Browning & Veit, 2022b). That is, given enough relevant evidence (of the kind described in the previous paragraph) we should assume that insects are sentient and treat them accordingly. This is because the moral costs of not treating them as sentient when they are in fact sentient are greater than the costs of treating them as sentient if they are not. Because it is not possible to independently evaluate the sentience of all insect species (there are well over a million of them!) we can be guided by what we do know about particular species and extrapolate from this to other species in the same taxonomic order (Birch, 2017). Given the relatively limited direct understanding of sentience in many individual *vertebrate* species, this approach should not be too problematic. It is also important that we remain flexible regarding the capacity for sentience at the different life stages of insects and to continue to update our knowledge and hence our legal systems based on the most recent research findings.

If we accept that most (perhaps all) adult insects plausibly have the capacity for sentience, and that sentience is an important criterion for moral and legal status, then it follows that we need to take the idea of including insects in animal welfare legislation seriously. I propose, therefore, that the New Zealand *Animal Welfare Act* should be amended to include all adult insects as 'animals'. Currently the New Zealand *Animal Welfare Act* excludes 'any animal in the pre-natal, pre-hatched, larval, or other such development stage' (s 2 (1)(d)(ii)). It is not clear at this point

Table 1 Evidence for sentience in selected invertebrate species

Taxonomic group	Number of sentience criteria met with high or very high confidence (out of 8)	Overall sentience evaluation
Molluscs [a]		
Octopuses*	7	Very strong evidence
Cuttlefish	4	Substantial evidence
Squid*	4	Substantial evidence
Crustaceans [a]		
True crabs*	5	Strong evidence
Anomuran crabs*	3	Substantial evidence
Astacid lobsters/ crayfish*	4	Substantial evidence
Spiny Lobsters*	2	Some evidence
Insects [b]		
Cockroaches	6	Strong evidence
Beetles	2	Some evidence
Flies	6	Strong evidence
Wasps, bees, ants	4	Substantial evidence
Moths, butterflies	3	Substantial evidence
Crickets, grasshoppers	3	Substantial evidence

*Invertebrate groups currently represented in the *Animal Welfare Act 1999* (NZ)
[a] Birch et al. (2021)
[b] Gibbons et al. (2022)

whether there is enough evidence for sentience in enough insect orders to repeal this exclusion for juveniles (see Gibbons et al., 2022), but future work should guide whether this also needs to be amended if the evidence warrants it.

The inclusion of insects could have a number of positive impacts on insect well-being and help to promote the broader cause of invertebrate justice. First, including insects would help to raise awareness about sentience in insects among the wider public which, in turn, might help to shape the way that people in general interact with these animals. The law here has a role in altering the normative landscape in which we inhabit

(Sunstein, 1996) and to overcome the many prejudices that people often display towards insects and other invertebrates (Fukano & Soga, 2021). Second, the inclusion of insects would enable contexts in which insects are reared in large quantities to be subject to clear and explicit welfare guidelines. In Aotearoa New Zealand, beekeeping is the industry that is most likely to be impacted. Indeed, there are over 700,000 commercial beehives in New Zealand containing around 50,000 bees each, so the scale of the impact of including bees in the *Animal Welfare Act* is substantial (Apiculture New Zealand, 2023). In many respects the aims of beekeepers and the well-being of bees converge, as illustrated in the New Zealand honey bee care code (Apiculture New Zealand, 2021). However, there will inevitably be contexts where welfare and commercial interests do not perfectly align and having explicit welfare guidelines will help to protect the interests of individual bees (Garrido & Nanetti, 2019). A third important benefit of including insects in the *Animal Welfare Act* is that research involving insects would need to be subject to institutional ethics review, as is the case with research on other animals currently included in the Act. This would cover critical issues such as the appropriateness of particular research protocols, the housing of animal subjects, the use of analgesics and anaesthetic, and appropriately humane methods of euthanasia. Although there is uncertainty regarding many of these areas, work is actively being undertaken that could guide best practice (Fischer et al., 2024).

It is, of course, not hard to envision various objections to the inclusion of insects in the *Animal Welfare Act*. For a start, many insect species are considered 'pests' in Aotearoa New Zealand, much as they are in other parts of the world. This includes insects that are threats to endemic wildlife (e.g. various introduced wasps), agriculture (e.g. aphids, white butterfly, codling moth), and human and animal well-being (e.g. lice, flies, mosquitoes). Many people also find insects annoying and there are large-scale commercial and residential pest control industries that focus on eliminating insect life. However, the existence of insect pests doesn't necessarily have to be problematic as there are also many vertebrate species that are labelled 'pests' in New Zealand such as possums, rats, and stoats and are treated in ways that have not impeded efforts to control their spread. A case can be made that we *should* treat these vertebrate species with greater compassion than is typically found in Aotearoa New Zealand (Wallach et al., 2018), however it suggests that the mere existence of insect 'pests' should not preclude their inclusion in the *Animal Welfare Act*.

A second potential objection is that many of our everyday actions—gardening, driving, walking—often result in inadvertent insect suffering and death. Given that animal welfare offences in New Zealand are strict liability offences, this could mean that many, if not all, people are potentially culpable of inflicting harm to insects. The inclusion of insects, however, wouldn't necessitate a Jain monk existence, gently sweeping the pavement in front of us to avoid stepping on them (Verchery, 2024). After all, many vertebrate species are also inadvertently killed in large numbers due to everyday activities like driving, without resulting in any increase in criminal convictions, highlighting the substantial prosecutorial discretion involved in enforcing animal welfare laws (Ferrere, 2022). It is not likely, therefore, that including insects would result in any kind of significant legal net widening. A final more general issue is that the inclusion of insects might be seen by some as a kind of *reductio ad absurdum* that could undermine the hard work of animal activists striving to make lives better for vertebrate species under the current *Animal Welfare Act* by leading the public to ignore welfare concerns for vertebrates (Durrant, 2024). However, given that the inclusion of neglected species like crustaceans in animal welfare laws has not led to an undermining of such laws it is not clear that extending legal protection to insects would lead to this outcome. Of course, whether the inclusion of insects would actually make a real difference in reducing harm and suffering remains an open question and serves as a reminder of the significant limitations of animal welfare laws more generally as they are currently formulated and administered.

Animal welfare laws have a number of widely recognised problems: they tend to be profoundly anthropocentric, riddled with exemptions and exceptions, and weakly enforced (Fasel & Butler, 2023). The existence of animal welfare laws in Aotearoa New Zealand and in other countries has done relatively little to stem the harm inflicted upon nonhuman animals. One major issue is that animal welfare and animal cruelty laws are consistently underenforced (e.g. Ferrere, 2022; Morton et al., 2020) with an unknown, but likely very large dark figure of offending. Although there has been no specific work on the topic, underenforcement is likely to be particularly problematic for less charismatic species like fishes and invertebrates. Indeed, although detailed information about convictions under the *Animal Welfare Act* in New Zealand is difficult to obtain, it is very probable that no one has been convicted of an offence against any invertebrate species under this Act. Ultimately, although current animal welfare legislation provides animals with, what Stucki (2020) terms 'simple animal rights', they fall far short of offering 'fundamental animal rights' of the kind that would recognise their intrinsic or ultimate value. The granting of all sentient animals fundamental rights of the kinds that humans (should) possess is arguably the best way to advance justice for vertebrate and invertebrate species alike (Cochrane, 2018; Durrant, 2024) and to help to reduce the harm that we inflict on non-human animals.

5 Conclusion

Most animal life on the planet is invertebrate life. Many invertebrate species quite possibly have the capacity for sentience: they can experience pleasures and pains. Humans, therefore, cause suffering and death to an extremely large number of individuals who have the capacity to subjectively experience the harms of our actions. Despite this, invertebrates have been, to date, relatively neglected in animal ethics, criminology, legal studies, and relevant legislation. In making a case for invertebrate justice I argue that we need to first and foremost recognise that we have moral and justice obligations towards sentient invertebrates and that we need to better include them in relevant laws and policies. More specifically, I argue that a case can be made that the New Zealand *Animal Welfare Act* should be amended to include all adult insects. This change has the potential to shape normative practices in relation to insect species in ways that can reduce harm and suffering. Offering legal protection to insects may also help to shift broader attitudes towards insects and other invertebrates in ways that may lead to a better recognition of the important role that invertebrates play in maintaining the integrity and flourishing of the ecosystems that all animals—including humans—depend upon. Ultimately the full realisation of invertebrate justice will depend on the recognition that all sentient beings, including invertebrates should have fundamental rights (Durrant, 2024).

References

Animal Welfare Act 1999 (NZ).

Animal Welfare (Sentience) Act 2022 (UK).

Andrews, K., Birch, J., & Sebo, J. (2025). Evaluating Animal Consciousness. *Science, 387*(6736), 822–824. https://doi.org/10.1126/science.adp4990

Apiculture New Zealand. (2021). New Zealand Bee Care Code. https://apinz.org.nz/faq-bee-care-code/

Apiculture New Zealand. (2023). *Thriving Together, Futureproofing New Zealand Apiculture: New Zealand Honey Strategy 2024–2030.*

Arkow, P. (2021). "Humane Criminology": An Inclusive Victimology Protecting Animals and People. *Social Sciences, 10*(9), 335. https://doi.org/10.3390/socsci10090335

Beirne, P. (1999). For a Nonspeciesist Criminology: Animal Abuse as an Object of Study. *Criminology, 37*, 117–147. https://doi.org/10.1111/j.1745-9125.1999.tb00481.x

Birch, J. (2017). Animal Sentience and the Precautionary Principle. *Animal Sentience, 16*(1). https://doi.org/10.51291/2377-7478.1200

Birch, J. (2024). *The Edge of Sentience: Risk and Precaution in Humans, Other Animals, and AI*. Oxford University Press.

Birch, J., Burn, C., Schnell, A., Browning, H., & Crump, A. (2021). *Review of the Evidence of Sentience in Cephalopod Molluscs and Decapod Crustaceans.* London School of Economics.

Blattner, C. E. (2019). The Recognition of Animal Sentience by the Law. *Journal of Animal Ethics, 9*(2), 121–136. https://doi.org/10.5406/janimalethics.9.2.0121

Blattner, C. E. (2024). The Nature and Scope of the Legal Recognition of Animal Sentience: A Functional Comparative Analysis of the Lex Lata, and Critical Remarks on its Future Development. In J. Kotzmann & M. B. Rodriguez Ferrere (Eds.), *The Legal Recognition of Animal Sentience: Principles, Approaches and Applications* (pp. 11–28). Hart Publishing.

Browning, H., & Veit, W. (2022a). Longtermism and Animals. In H. Greaves, D. Thorstad, & J. Barrett (Eds.), *Essays on Longtermism*. Oxford University Press.

Browning, H., & Veit, W. (2022b). The Sentience Shift in Animal Research. *The New Bioethics, 28*(4), 299–314. https://doi.org/10.1080/20502877.2022.2077681

Brusca, R. C., & Brusca, G. J. (2003). *Invertebrates* (2nd Ed.). Sinauer Associates Limited.

Cardoso, P., Barton, P. S., Birkhofer, K., & Samways, M. J. (2020). Scientists' Warning to Humanity on Insect Extinctions. *Biological Conservation, 242*. Article 108426. https://doi.org/10.1016/j.biocon.2020.108426

Cavan, E., Grilly, E., Reig, K., & Mackay, N. (2022). *Antarctic Krill: Powerhouse of the Southern Ocean*. WWF-Australia.

Celermajer, D., Chatterjee, S., Cochrane, A., Fishel, S., Neimanis, A., O'Brien, A., & Waldow, A. (2020). Justice Through a Multispecies Lens. *Contemporary Political Theory, 19*(3), 475–512. https://doi.org/10.1057/s41296-020-00386-5

Cochrane, A. (2018). *Sentientist Politics: A Theory of Global Interspecies Justice*. Oxford University Press.

Cripps, E. (2022). *What Climate Justice Means and Why We Should Care*. Bloomsbury.

Donaldson, S., & Kymlicka, W. (2011). *Zoopolis: A Political Theory of Animal Rights*. Oxford University Press.

Durrant, R. (2024). *Invertebrate Justice: Extending the Boundaries of Non-Speciesist Green Criminology*. Palgrave.

Eisenhauer, N., & Hines, J. (2021). Invertebrate Biodiversity and Conservation. *Current Biology, 31*(19), R1214–R1218.

Fasel, R. N., & Butler, S. C. (2023). *Animal Rights Law*. Hart Publishing.

Ferrere, M. B. R. (2022). Animal Welfare Underenforcement as a Rule of Law Problem. *Animals: An Open Access Journal from MDPI*, *12*(11), Article 1411. https://doi.org/10.3390/ani12111411

Ferrere, M. B. R. (2024). Legal Recognition of Sentience in New Zealand. In J. Kotzmann & M. B. Rodriguez Ferrere (Eds.), *The Legal Recognition of Animal Sentience: Principles, Approaches and Applications* (pp. 155–171). Hart Publishing.

Fischer, B., Barrett, M., Adcock, S., Barron, A., Browning, H., Chittka, L., Crump, A., Drinkwater, E., Gibbons, M., Haverkamp, A., & Perl, C. (2024). *Guidelines for Protecting and Promoting Insect Welfare in Research*.

Insect Welfare Research Society. https://static1.squarespace.com/static/633dcce66ff3284353256a63/t/65ad7a8231bdea211355ec83/1705867906190/IWRS+Guidelines.2024.Insect.pdf

Fukano, Y., & Soga, M. (2021). Why Do So Many Modern People Hate Insects? The Urbanisation-Disgust Hypothesis. *Science of the Total Environment, 777*, Article 146229. https://doi.org/10.1016/j.scitotenv.2021.146229

Garrido, C., & Nanetti, A. (2019). Welfare of Managed Honey Bees. In C. Carere & J. Mather (Eds.), *The Welfare of Invertebrate Animals* (pp. 69–104). Springer.

Gibbons, M., Crump, A., Barrett, M., Sarlak, S., Birch, J., & Chittka, L. (2022). Can Insects Feel Pain? A Review of the Neural and Behavioural Evidence. *Advances in Insect Physiology, 63*, 155–229. https://doi.org/10.1016/bs.aiip.2022.10.001

Girgen, J. (2022). Neglecting Nonhuman Animals: A Content Analysis of Introductory Criminal Justice and Criminology Textbooks. *Society and Animals, 30*, 569–588. https://doi.org/10.1163/15685306-BJA10009

Goulson, D. (2021). *Silent Earth: Averting the Insect Apocalypse*. Random House.

Herzog, H. (2021). *Some We Love, Some We Hate, Some We Eat: Why It's So Hard to Think Straight About Animals* (2nd Ed.). Harper.

Kotzmann, J. (2023a). A Typology of Australian Animal Sentience Recognition Provisions—Enacted and Proposed. *Federal Law Review, 51*, 157–181. https://doi.org/10.1177/0067205X231166707

Kotzmann, J. (2023b). Sentience and Intrinsic Worth as a Pluralist Foundation for Fundamental Animal Rights. *Oxford Journal of Legal Studies, 43*, 405–428. https://doi.org/10.1093/ojls/gqad003

Kotzmann, J. (2024). Sentience Recognition as a Means for Moving Beyond Animal Welfare to Animal Rights. In J. Kotzmann & M. B. Rodriguez Ferrere (Eds.), *The Legal Recognition of Animal Sentience: Principles, Approaches and Applications* (pp. 63–78). Hart Publishing.

Kotzmann, J., & Ferrere, M. B. R. (Eds.), (2024). *Principles, Approaches and Applications*. Hart Publishing.

Mikhalevich, I., & Powell, R. (2020). Minds Without Spines: Evolutionarily Inclusive Animal Ethics. *Animal Sentience, 29*(1). https://doi.org/10.51291/2377-7478.1527

Milkman, R., Luce, S., & Lewis, P. (2014). Occupy Wall Street. In J. Goodwin & J. M. Jasper (Eds.), *The Social Movements Reader* (pp. 30–44). Wiley Blackwell.

Morton, R., Hebart, M. L., & Whittaker, A. L. (2020). Explaining the Gap Between the Ambitious Goals and Practical Reality of Animal Welfare Law Enforcement: A Review of the Enforcement Gap in Australia. *Animals, 10*(3) Article 482. https://doi.org/10.3390/ani10030482

Nurse, A. (2021). Green Criminological Perspectives on Dog-Fighting as Organised Masculinities-Based Animal Harm. *Trends in Organized Crime, 24*(4), 447–466. https://doi.org/10.1007/s12117-021-09432-z

Perry, M. W. (2024). Why Sentience Should be the Only Basis of Moral Status. *The Journal of Ethics, 28*(4), 719–741. https://doi.org/10.1007/s10892-024-09487-4

Piketty, T. (2024). *Nature, Culture, and Inequality*. Scribe Publications.

Pons-Hernandez, M. (2024). "Missing the Trees for the Forest?" An Analysis of the Harms to European Eels Caused by Their Trafficking and Trade. *Critical Criminology, 32*(1), 77–95. https://doi.org/10.1007/s10612-024-09765-5

Regan, T. (2004). *The Case for Animal Rights* (2nd Ed.). University of California Press.

Rowe, A. (2020). The Scale of Direct Human Impact on Invertebrates. EA Forum. https://forum.effectivealtruism.org/posts/9drbh8sKzzykaX38P/the-scale-of-direct-human-impact-on-invertebrates

Rowlands, M. (2025). *Animal Rights*. MIT Press.

Sandler, R. (2018). *Environmental Ethics: Theory in Practice*. Oxford University Press.

Singer, P. (2023). *Animal Liberation Now*. The Bodley Head.

Smith, M. R., Mueller, N. D., Springmann, M., Sulser, T. B., Garibaldi, L. A., Gerber, J., & Myers, S. S. (2022). Pollinator Deficits, Food Consumption, and Consequences for Human Health: A Modeling Study. *Environmental Health Perspectives, 130*(12), Article 127003.

Sollund, R. (2021). Nonspeciesist Criminology, Wildlife Trade, and Animal Victimization. In H. N. Pontell (Ed.), *Oxford Research Encyclopaedia of Criminology*. https://doi.org/10.1093/acrefore/9780190264079.013.608

Sollund, R. (2024). Anthropocentrism, Speciesism and Speciecide. In *Criminological Connections, Directions, Horizons* (pp. 116–132). Routledge.

Stork, N. E. (2018). How Many Species of Insects and Other Terrestrial Arthropods are there on Earth? *Annual Review of Entomology, 63*, 31–45. https://doi.org/10.1146/annurev-ento-020117-043348

Stucki, S. (2020). Towards a Theory of Legal Animal Rights: Simple and Fundamental Rights. *Oxford Journal of Legal Studies, 40*, 533–560. https://doi.org/10.1093/ojls/gqaa007

Sunstein, C. R. (1996). On the Expressive Function of Law. *University of Pennsylvania Law Review, 144*(5), 2021–2053. https://doi.org/10.2307/3312647

Taylor, N., & Fitzgerald, A. (2018). Understanding Animal (Ab)use: Green Criminological Contributions, Missed Opportunities and a Way Forward. *Theoretical Criminology, 22*(3), 402–425. https://doi.org/10.1177/1362480618787173

Trebilco, R., Melbourne-Thomas, J., & Constable, A. J. (2020). The Policy Relevance of the Southern Ocean Food Web Structure: Implications of Food Web Change for Fisheries, Conservation and Carbon Sequestration. *Marine Policy, 115*, Article 103832. https://doi.org/10.1016/j.marpol.2020.103832

Verchery, L. (2024). Learning to Walk Softly: Intersecting Insect Lifeworlds in Everyday Buddhist Monastic Life. In D. Aftandilian (Ed.), *Animals and Religion* (pp. 107–111). Routledge.

Wallach, A. D., Bekoff, M., Batavia, C., Nelson, M. P., & Ramp, D. (2018). Summoning Compassion to Address the Challenges of Conservation. *Conservation Biology, 32*(6), 1255–1265. https://doi.org/10.1111/cobi.13126

Walters, R., & Couper, A. (2023). Political Economy and the Government Attack on Sharks: A Non-Speciesist Southern Green Criminology. In D. R. Goyes (Ed.) *Green Crime in the Global South: Essays on Southern Green Criminology* (pp. 141–162). Springer International.

White, R. (2013). *Environmental Harm: An Eco-Justice Perspective.* Policy Press.

White, R. (2018). Green Victimology and Non-Human Victims. *International Review of Victimology, 24*(2), 239–255. https://doi.org/10.1177/0269758017745615

Wyatt, T. (2024). Criminology's Animal Turn. In E. Carrabine & A. Di Ronco (Eds.), *Criminological Connections, Directions, Horizons* (pp. 100–115). Routledge.

10

Justice Through Deterrence: Media Framings of the Rhino Horn Trade in Vietnam

Michael Scott Smith

1 Introduction

A rhino poaching crisis in South Africa, driven in part by demand from Vietnam, has persisted for 17 years. Between 2008 and 2016, despite peak poaching years, Vietnam reported no prosecutions (Worldwide Fund for Nature (WWF), 2016a), creating a 'low-risk, high-profit' environment for traffickers (Scanlon, 2016). As global pressure mounted for stronger enforcement of illegal wildlife trade ('IWT'), Vietnam began taking action towards greater criminalisation and enforcement, including through revisions to the Penal Code that took effect in 2018. This chapter explores the narratives used to portray the Vietnamese rhino horn trade using both press releases sent to Vietnamese journalists from environmental Non-Government Organisations (NGOs) and articles in Vietnamese digital newspapers. The findings demonstrate the impact

M. S. Smith (✉)
Griffith Criminology Institute, Brisbane, QLD, Australia
e-mail: m.smith@griffith.edu.au

R. Killean et al. (eds.), *Conceptualising Justice*, Palgrave Studies in Green Criminology, https://doi.org/10.1007/978-3-032-19791-7_10

that international stakeholder and NGO pressure through local Vietnamese media appears to have played on enforcement of laws against the rhino horn trade in Vietnam.

This chapter, and the broader scope of this research, are underpinned by the concept of deterrence (Braga & Weisburd, 2012). Whether punishment deters by making the criminal pay for the harm they caused or by incapacitating them, the principle of deterrence rests on society's right to be protected from crime—or in this case on rhinos being 'deemed to have the right to be protected' (Loughran et al., 2016: 50). Within this framework, justice for rhinos means protecting them from further harm, exploitation, and suffering. Deterrence, brought about by the perception of celerity, certainty and severity of punishment for perpetrators of the rhino horn trade, arguably aligns with both anthropocentric and eco-centric worldviews. Meaning, it both understands the rhino trade as an affront to *human* perceptions of harm and justice (anthropocentric) and emphasises the intrinsic worth of rhinos by advocating for crimes against them to be punished (eco-centric). From a green criminological perspective of deterrence, the international pressure on Vietnam to enforce laws against the rhino horn trade functions as a form of preventative justice for them.

However, not all actors engaged in the discourse concerning rhino protection share the same view of justice. As this chapter illustrates, although international advocacy—including from conservation organisations—appears to have prompted Vietnam to take a firmer stance against the illegal rhino horn trade, media portrayals in Vietnamese newspapers remain largely human-centric, framing rhino poaching as a crime without engaging in the inherent worth of rhinos or their right to be free from suffering. Although media coverage can be seen as a positive step in highlighting the plight of rhinos, prevailing narratives largely mirror external pressures that drive criminal justice responses. Advancing justice for rhinos, I argue, may be more meaningfully achieved through media portrayals which recognise their inherent rights.

This chapter draws on my broader research agenda into the rhino horn trade (Smith, 2022, 2024, 2025) using green criminology. Green criminology is 'a perspective within criminology' that seeks to 'identify environmental crimes and non-statutorily proscribed environmental harms

– to investigate their causes and consequences, as well as the possible responses thereto' (Brisman, 2018: 469). The rhino horn trade is difficult to quantify and understand because of its clandestine nature, however discrete seizure events expose it as an ecological harm that has clearly identifiable actors, non-human victims and recognisable injuries. This renders it a useful case study in green criminology. My research sits more specifically at the intersection of green criminology and media representations, known as 'green-cultural criminology' (Brisman & South, 2014). A cultural criminology approach concerns the 'social construction of green harms, particularly the media's construction of environmental harms, and the political dynamics of constructing environmental harms' (Lynch et al., 2021: 360). This chapter draws from this approach to explore how rhino horn trafficking is constructed and framed in the media.

2 The Rhino Horn Trade in Vietnam

Between 2012 and 2024, more than 11,000 rhinos were poached in Africa (Save the Rhino n.d.). The wildlife trade monitoring NGO TRAFFIC identified Vietnam as a nexus for the illegal trade (Milliken & Shaw, 2012), and by 2021 the Environmental Investigation Agency (EIA) still named Vietnam as 'one of the Parties most affected by rhinoceros horn trafficking' (EIA, 2021: 8). Vietnam's strong culture of wildlife consumption (Drury, 2011; Van Song, 2008) has fuelled demand, prompting environmental NGOs to intervene, with TRAFFIC publishing a comprehensive report on the rhino horn trade as early as 2012 (Milliken & Shaw, 2012). The lucrative nature of the transnational trade, combined with 'the ease with which items can be globally transported' (Wellsmith, 2011: 136) and the low risk of detection and punishment (Scanlon, 2016), has made rhino horn trafficking particularly attractive to organised crime (McMullan and Perrier in Wellsmith, 2011: 145).

A shifting and increasingly complex geopolitical ecology of conservation is emerging around the world, characterised by growing efforts to frame the IWT not merely as a conservation issue, but as a serious crime and a matter of national and international security. As noted by Massé et al. (2020), IWT has transitioned from being primarily managed by conservation actors to being addressed through the lens of crime and security, involving law enforcement, intelligence agencies, and judicial systems. Enforcement elites have justified this shift by framing IWT as being run by transnational, organised, sophisticated, and ruthless groups of criminals (Massé et al., 2020). This transformation has led to a policy environment that prioritises legal and judicial reforms, criminal investigations, surveillance, and enforcement technologies (Massé & Margulies, 2020). In doing so, policy often sidelines alternative approaches that focus on addressing the root causes of IWT, such as poverty, lack of livelihood opportunities, and consumer demand (Massé & Margulies, 2020).

In line with this geopolitical shift, international and civil society have intensified pressure on Vietnam regarding its lack of commitment and determination to fight the rhino horn trade. In the main, this pressure has had a serious crime rhetoric (Smith, 2025). As a result of increased pressure, including threats of trade sanctions at a Convention on International Trade in Endangered Species of Wild Fauna and Flora ('CITES') conference in October 2016, the Vietnam government officially made a commitment to fight IWT through the Hanoi Statement (2016). Although CITES can 'create an illusion that wildlife trade is subject to control and enforcement that are actually lacking' (Sollund, 2022: 1019), and the extent of enforcement in Vietnam may differ from its public pronouncements, there is evidence that Vietnam has begun to take the IWT more seriously. It has revised its 2015 Penal Code for rhino horn trade perpetrators, with revisions coming into effect on 1 January 2018. The amended penal code states that convictions for rhino horn traffickers carry prison sentences up to 15 years. The revised legal framework aligns with serious crime definitions of IWT as per the UN Convention against Transnational Organized Crime ('UNTOC') under the United Nations Office on Drugs and Crime ('UNODC'). Under this convention, serious crimes such as drugs and weapons trafficking and money laundering attract a minimum penalty of four years imprisonment (UNODC, 2004).

The first ever destruction of Vietnam's rhino horn and ivory stockpiles was witnessed by representatives of 42 countries in November 2016 (Lever, 2017). This event was intended to symbolise Vietnam's commitment and determination to fight IWT. From 2016 onwards Vietnam began prosecuting more individuals arrested at seizures, and delivering prison sentences in line with its CITES commitments. The rate of arrests compared to prosecutions seemed to improve from 2018 onwards. From 2018 to September 2021 the direct-action environmental NGO, Education for Nature Vietnam (ENV) published the Wildlife Crime Unit Incident Tracking Database which recorded 20 rhino horn trafficking cases that resulted in an arrest, 14 of which were prosecuted and 11 of which received serious sentences of more than four years' imprisonment (ENV Facebook, pers comm, 24 February 2021). In the following sections, I explore what news framing reveals about these developments.

3 News Framing as Methodology

Public education can play an important role in strengthening law enforcement and enhancing deterrence. For this to be effective, however, communication about laws must reach not only citizens but also law enforcement officers and government officials, particularly where laws are poorly enforced or ambiguous (Goyes & Sollund, 2016: 90). The news media can contribute by shaping awareness and attitudes towards IWT (Wyatt, 2014). Given that one of the persistent challenges in IWT enforcement is its limited deterrent effect (Wellsmith, 2011), enforcement strategies need to be informed by the public's knowledge and experience of these activities. This is especially relevant in Vietnam, where laws against the rhino horn trade have often lacked clarity and consistent enforcement.

News framing is not only a way for journalists to efficiently convey information to audiences (Gitlin, 1980), but also a persuasive tool for political communicators, who often use media debates to influence public opinion and policy (Pan & Kosicki, 2001). Framing is a process which involves selecting *'some aspects of a perceived reality and make them more salient in a communicating text, in such a way to promote a particular problem definition, causal interpretation, moral evaluation and/or treatment recommendation* for the item described' (Entman, 1993: 52 emphasis in original, see also Gitlin, 1980). Words, phrases, exemplars, images and metaphors can each act as framing devices in news media (Gamson & Modigliani, 1989; Van Gorp, 2007).

Focusing on news framing can lead to greater understanding of how environmental issues are politically deliberated (Hansen, 2011). Effective advocacy requires strategic framing to gain legitimacy and prompt action (Gitlin, 1980; Solesbury, 1976). Domestically, the media plays a key role in raising awareness among the public and policymakers (Spector & Korula, 1993), while at a global level media attention may motivate countries to reduce environmental harms (Mitchell, 2003; Tallberg, 2002) and sell the problem to policymakers (Garstecki, 2006). Furthermore, media attention may enhance understanding and promote the adoption of international environmental agreements (Spector & Korula, 1993). When combined with international pressure, such attention can make compliance with international agreements more attractive (Mitchell, 2003; Tallberg, 2002). Finally, transnational political communication, facilitated by supranational organisations, international elites and local media, helps sustain global attention and action on environmental issues (Fraser, 2007; Lester, 2014).

In Vietnam, different actors are able to influence news frames. The government influences news frames, placing pressure on Vietnamese news organisations to print propaganda in relation to international and domestic legal developments (Pham et al., 2017). However, Vietnamese official sources are often joined by environmental NGO and international sources (Pham et al., 2017) in news reports. This suggests there may be a growing green public sphere, as has been similarly theorised in the Chinese context following increased access to media amongst environmental groups (Yang & Calhoun, 2013). International influence can also be achieved through news framing. In the context of encouraging Vietnam's compliance with CITES, Knupfer and Entman (2018) argue that new international interpretive communities comprising nation states and political elites have used framing to shape foreign political decision-making and institutions. Such global-level framing may be 'aimed at localised institutional contexts' (Knupfer & Entman, 2018: 484).

The chapter draws on framing theory to explore media representations of the rhino horn trade. To do so, it combines two data sets. The first is comprised of collated news frames drawn from environmental NGO press releases between 2014 and 2016. These press releases were sent out to Vietnamese journalists by seven environmental NGOs with offices in Vietnam as part of their campaigns to reduce demand for rhino horn (Smith, 2022). Deeper understanding of these specific frames stemmed from semi-structured interviews with 11 key NGO personnel, three Vietnamese journalists, a newspaper editor and five experts in Vietnamese journalism. Most of the interviews with NGO personnel were conducted on site at the 2016 Hanoi IWT conference.

The second dataset was collected as an extension project which looked more closely at the use of criminality framing. It consists of 144 news articles appearing in four Vietnamese online newspapers during a sample period of 2016 to 2020 (Smith, 2024). The sampling period was designed to include the 2016 Hanoi IWT Conference as this was an event where Vietnam sought to emphasise their commitment and determination to fight IWT including the rhino horn trade. The articles were translated from Vietnamese into English using machine translation, a method that has dramatically improved in accuracy since first introduced (High, 2023) and is well suited for analysis of news articles (Turner et al., 2014: 2). To ensure accuracy, a component of human translation and post-editing was incorporated (Turner et al., 2014: 525). This was followed by semi-structured interviews with 10 Vietnamese journalists, three NGO personnel and an expert on Vietnamese media.

By collating NGO press release frames and interrogating criminality frames in Vietnamese digital newspapers, this chapter aims to provide a nuanced picture of the reporting of a transnational environmental crime, giving a unique insight into Vietnamese justice from a green criminological perspective. Analysis of both data sets was carried out inductively by a close reading of the printed-out material including text, photographs and graphics. Words and sections were underlined in pencil and notes about the themes that emerged were taken in the margins. Patterns and themes were interpreted through framing devices in the text and images. Once the key frames were identified and classified, they were discussed with my PhD supervisors. Then in an analogue process, framing devices which signified respective frames were highlighted in different colours. The frames were then described, broken down into specific aspects and their framing devices tabulated. The following examples illustrate how frames were identified. Words and phrases that indicated that Vietnam was responsible for the rhino poaching crisis in Africa signified the Responsibility frame; while words and phrases indicating organised criminal networks, such as 'mafia', 'boss' or 'chain' were highlighted as signifying the Serious Crime frame. Words and phrases such as 'committed' and 'determined' signified the Compliance frame. These framings are explored in the following section.

4 Framing the Rhino Horn Trade

Analysis of NGO press releases and Vietnamese news articles indicates that a criminality-focused framing of the rhino horn trade persisted even as NGOs and international actors sought to pressure or encourage the Vietnamese government to take stronger action. I identified several frames pressuring or encouraging Vietnam to strengthen and enforce criminal laws against the rhino horn trade, including Responsibility, Brutality, Lip Service (Smith, 2022), Serious Crime, Compliance and Lack of Enforcement (Smith, 2024, 2025). I discovered Responsibility and Lip Service together in the NGO press releases. NGOs stressed that demand for and illegal trade in rhino horn in Vietnam drives the poaching crisis in Africa. Yet they argued that, in practice, Vietnam has paid only lip service to combating the trade, offering rhetoric without meaningful action. In contrast, digital newspapers' framing focused on the rhino horn trade as a 'Serious Crime', evidence of and claims of Vietnam's 'Compliance' with its CITES commitments and an ongoing Lack of Enforcement.

4.1 Responsibility and Lip Service

Between 2014 and 2016, NGOs enacted four key frames aimed at reducing demand for rhino horn in their press releases: Responsibility,

Lip Service, Voodoo Wildlife Parts (a science frame) and Empowering Youth (Smith, 2022: 118). Of these, Responsibility and Lip Service both related strongly to concepts of justice and are considered here. A sub-frame of Responsibility, Brutality, communicated that Vietnam was responsible for the brutal nature of the poaching attacks that occurred in Africa.

The Responsibility frame conveyed that 'Vietnamese rhino horn consumers are responsible for driving the poaching in Africa' (Smith, 2022: 118). The frame, which was present in 68% of the sampled NGO press releases (Smith, 2022: 118), drew on the 'attribution of responsibility' model (Dirikx & Gelders, 2009; Semetko & Valkenburg, 2000), connecting Vietnam's responsibility to consequences such as rhino poaching statistics, the brutality of killings in Africa, and especially the conservation crisis facing rhinos in South Africa. NGOs also used the poaching of Vietnam's last native rhino in 2010 as an illustrative example of Vietnam's responsibility. Some press releases contained text and graphic images that stressed the brutal nature of poaching attacks (for an example see Smith, 2022: 117).

The Lip Service frame criticised the Vietnamese government for 'its poor record of enforcement of laws against illegal trade in rhino horn' (Smith, 2022: 118) and for failing to fulfil its international commitments under CITES. Quotes that enacted this frame tended to diplomatically note what Vietnam had done, before stating that there was more to do. This frame was present in 49% of the 76 press releases. Supporting this narrative, one NGO representative interviewed stated that Vietnam remained 'really weak' on enforcement and prosecution (Smith, 2024: 64) despite making positive progress reports (WWF, 2019). The press releases displaying these frames held Vietnam responsible and advocated for more meaningful action against the rhino horn trade while drawing attention to the country's international spotlight. The following excerpt from a press release illustrates the Lip Service frame:

> Vietnam's decision to host the conference [2016 Hanoi IWT Conference] is admirable and it has helped shine a much-needed light on the illegal wildlife trade across the Greater Mekong Region – a trade that is emptying the forests of wildlife and impacting species such as rhinos,

> elephants and pangolins in Africa. But … Vietnam needed to commit to more concrete action plans that will have an impact on the ground. It did pledge to strictly monitor domestic markets and eradicate illegal wildlife trade points, strengthen law enforcement, improve cross border cooperation, but much more detail is needed, especially on legislative reform. (WWF, 2016a)

Importantly, the Responsibility frame incorporated green criminological elements. For example: WildAid's campaign 'When the buying stops the killing can too' drew attention to the plight of rhinos (Smith, 2022: 120); some releases encouraged increased personal responsibility for the brutality inflicted on rhinos; while others highlighted the emotional impact of rhino horn consumption, such as leaving baby rhinos orphaned (Smith, 2022: 124). These sub-frames portray poaching as a violation of rhinos' rights, rejecting their commodification. However, the Lip Service frame, through its emphasis on detection, prosecution, and sentencing of actors involved in the rhino horn trade centred on criminality rather than rights violations.

4.2 Serious Crime, Compliance and Lack of Enforcement

My framing analysis of news articles revealed three key criminality-oriented frames: Serious Crime, Compliance and Lack of Enforcement (Smith, 2025). These frames were promoted through quotes and messaging from Vietnamese officials, international elites and NGO sources. Among them, Serious Crime and Compliance emerged as the most dominant. Although less prevalent, the Lack of Enforcement frame was closely related to the Compliance frame because it could easily be interpreted as a 'Lack of Compliance' narrative. This indicated systemic shortcomings in Vietnam's law enforcement, customs, and courts in addressing the rhino horn trade.

When Vietnam adopted the UNODC serious crime definition for rhino horn trafficking it was under additional pressure to act. The UN Resident Coordinator in Vietnam, Pratibha Mehta, framed the IWT as an 'extremely dangerous issue' (Voung, 2016), illustrating the Serious

Crime frame. News sources further framed rhino poaching as a Serious Crime by: depicting the involvement of three or more in the smuggling chain in an organised criminal network (Smith, 2025); grouping IWT with other serious transnational crimes (Smith, 2025); and likening IWT syndicate members such as the 'Boonchai Bach' to the mafia or the 'Corleones' from the movie *The Godfather* (Tuong, 2018). Additional Serious Crime framing devices included a focus on organised criminal activity such as corruption (Smith, 2025); depictions of syndicates as ruthless or sophisticated (Smith, 2025) and descriptions of rhino horn smuggling or IWT as a 'national security' (Ngoc, 2018), health security (Huong, 2020) or economic security issue (Voung, 2016).

Illustrating the pressure on Vietnam to comply with international expectations in the lead-up to the 2016 Hanoi IWT conference, the digital arm of the *Cong An Nhan Dan* newspaper (Vy, 2016) reported that the international community was threatening sanctions if Vietnam failed to act:

> The World Wide Fund for Nature (WWF) has just sent out a message calling on Vietnam to be more aggressive with the illegal rhino horn trade, which is still going strong, otherwise it will face serious threats and sanction orders.

Soon after the WWF statement, Vietnam appeared to respond to these 'threats' from the major international conservation player by presenting Hanoi's IWT conference as a symbol of its commitment and determination to fight IWT. Moreover, an event on the sidelines of the conference was covered extensively in all four selected digital newspapers in Vietnam. The newspapers published photo essays that featured the burning or crushing of rhino horn, thus safeguarding it from ever re-entering the trade. These photo essays also highlighted a heavy law enforcement presence alongside numerous international stakeholders and NGO representatives, visually signalling security and projecting Vietnam's strong compliance with its international commitments.

In *Lao Dong* (16 December 2019) the visiting South African ambassador Mr Lekgoro was quoted placing pressure on Vietnam. He stated that it was a 'primary concern' of South Africa to stop rhino poaching and for Vietnam to cooperate at an international level to fight the trade. In the Vietnamese digital newspapers Vietnam responded to international pressure with acts and statements of compliance. The excerpt below describes a MoU that committed Vietnam to cooperate with South Africa to prevent crimes against rhinos (Le Quang, 2019):

> Since Vietnam and South Africa signed a Memorandum of Understanding on biodiversity conservation and protection to date, this is the 4th time that Vietnam has handed over rhino horn inspection specimens confiscated from arrests to South Africa - in order to expertly determine the origin and serve the effective criminal investigation. The handover of this specimen demonstrates Vietnam's strong commitment to strictly and fully implement international commitments to which Vietnam is a member; At the same time, it shows the good cooperation between the two countries in the prevention and fight against wildlife crimes.

Vietnam's 'prevention and fight against wildlife crimes' suggests protecting rhinos from further harm provides them justice. Thus, it is a bridge between criminal justice approaches and green criminological perspectives. The Compliance frame was enacted by claiming or showing that Vietnam was toughening up on rhino horn smuggling. It reflects deterrence narratives that emphasise increased policing and the certainty of punishment, including the deployment of dedicated law enforcement personnel coordinated with agencies at both local and international levels.

In addition to deterring rhino horn trafficking through international cooperation, the news points to the deterrent effect of increased penalties (Ngoc, 2018): 'It [the Revised Penal Code 2015] has helped shape the penalty framework, to increase the severity of criminal penalties and increase deterrence'. Deterrence may be the intention of severe punishment, but the certainty of punishment has been shown to be a more effective deterrent (Nagin, 2013; Nagin & Pogarsky, 2001, 2003). Moreover, the media's increased focus on seizures could be viewed by the public as increasing the certainty of punishment or increasing the 'certainty of apprehension of punishment' (Nagin, 2018: 157). For example, Ngoc (2018) reported that more than 40 prison sentences for IWT offences were handed out between October 2016 and December

2017, including three sentences of imprisonment of three to seven years for ivory or rhino horn offences.

The concept of deterrence also relates to the Lack of Enforcement frame. In the years following 2016 several rhino horn traffickers received lengthy prison sentences that approached the 15-year maximum under the revised penal code. For example, a resident of Hanoi was sentenced to 12 years imprisonment in 2020 for 'illegally transporting 11 rhino horns' from Mozambique (Nhu, 2020). However, other sentencing decisions were criticised. In 2018 *Lao Dong* newspaper reported on a court appeal by a local environmental organisation, ENV. Quoting ENV's press release it criticised a 13-month sentence and VND10 million (US$500) fine as too lenient thus enacting Lack of Enforcement. They appealed because the convicted man was a leader of a group of traffickers, which denotes serious organised crime. The police had seized 23 rhino horns from the perpetrator's home. ENV claimed deterrence should be a key purpose of sentencing in this case (Thien, 2018):

> The penalty of 13 months imprisonment for defendant Nguyen Mau Chien is incompatible with the role of the defendant who leads, solicits and directs the whole process of transporting banned goods as well as storing banned goods of co-defendants and accomplices; There is no purpose to deter and prevent crimes on wildlife.

ENV's appeal states that a more severe sentence would be more compatible with the crime. This reflects green criminologist Sollund's (2022: 1030) argument that punishment needs to be proportional to the harm committed to non-humans. According to Sollund's view, justice would be served by sentencing kingpins, such as Nguyen Mau Chien, with more severe penalties. To further frustrate matters Chien's convictions were for 'transporting' and 'storing banned goods', a framing of the crime that reduces rhino horns to objects, minimising the harm to animals.

The following excerpt from a Vietnamese investigative journalist describes how, despite providing police with evidence of IWT activity, no action was taken. This is an example of the 'Lack of Enforcement' frame. Such an account further illustrates the pressure for stronger responses, underscoring that effective enforcement and sentencing are intended to achieve deterrence. As the journalist observes, the severity and certainty of punishment are 'not enough to deter' (Lang, 2019):

> The arrests and changes in the Penal Code are still not enough to deter illegal wildlife traders. As for the authorities, have they properly implemented their responsibilities in combating wildlife trafficking? In many localities, we have sent photos, information and addresses to investigators of the Environmental Police [who investigate IWT], but most of them gave no response.

International and NGO pressure was reflected in both the types of sources quoted in news stories and the frames they advanced. Across the 144 articles analysed, 77 unique sources were cited: by 20 international actors, 18 environmental NGOs, and 37 Vietnamese officials. As frames are closely tied to their communicators (Hertog & McLeod, 2001; Wichgers et al., 2021), these voices combined to reinforce a criminality frame that conveyed international pressure on Vietnam to take the rhino horn trade more seriously. Prominent international actors included Britain, the US, and South Africa. From Britain, elite sources such as Prince William and the British Ambassador were quoted, while the United Nations, CITES, US organisations, and South African ambassadorial and ministerial representatives also featured. International environmental NGOs such as WWF, the Wildlife Conservation Society (WCS), and TRAFFIC were frequently cited, alongside Vietnam's domestic enforcement-focused NGO, ENV.

The semi-structured interviews revealed potential reasons that international and NGO sources were able to draw media attention to the rhino trade in the Vietnamese digital newspapers. Vietnamese journalists tended to support rhino conservation and tough jail sentences for rhino horn traffickers (Smith, 2024). They were often under-resourced and financially incentivised by word count rather than time spent. Hence, they relied strongly on press releases whether from international sources, environmental NGOs or Vietnamese officials. Sometimes journalists copied and pasted the entire text or extracted quotes and information but may not have sought a balance of views. These journalistic constraints and routines might explain how international sources and NGOs were able to get their views and frames into the news, thus applying pressure through the media.

5 The Impact of News Framing in Vietnam

The framing of law enforcement against the rhino horn trade in Vietnamese digital newspapers appeared to have an 'impact on social policy' in Vietnam, playing a positive 'role in social control' (Hertog & McLeod, 2001: 160). Environmental NGOs attributed responsibility to Vietnam for the rhino horn trade and the rhino poaching crisis in Africa (Smith, 2022). The attribution of responsibility (Semetko & Valkenburg, 2000) was linked to the consequences (Dirikx & Gelders, 2009), with news stories blaming Vietnam for the harm and suffering to rhinos. Alongside calling Vietnam responsible in press releases, the Lip Service frame directed NGOs' frustration at Vietnam's actual lack of compliance with its CITES justice commitments. The Lip Service frame that NGOs enacted used the 'accountability/governance' model (Nisbet, 2009). Criticism inherent in the frame reinforced the view that IWT was not treated seriously (Wellsmith, 2011; Wyatt, 2014) and not given the priority it deserved (EIA 2013; WWF, 2016a, 2016b). To resolve this transnational conflict international environmental organisations such as TRAFFIC, WWF and WCS combined with nation states and political elites in mediated transnational political communication (Lester, 2014) 'aimed at localised institutional contexts' (Knupfer & Entman, 2018: 484). In the context of a globalised public sphere, these supranational sources successfully gained the attention of the local media (Papastergiadis, 2000), putting pressure on enforcement authorities to act.

However, NGO press releases rarely invoked the inherent rights of animals, whether human or non-human. This omission is perhaps unsurprising, as institutions such as CITES prioritise regulating wildlife trade over advancing the conservation and rights of all species (Sollund, 2022: 1029). Instead, NGO messaging centred on reducing demand and ensuring law enforcement. A pivotal moment came in 2016, when international stakeholders and NGOs intensified pressure on Vietnam to strengthen enforcement, particularly during the CITES Conference of the Parties (CoP). At this meeting, Vietnam faced sharp criticism and the threat of economic sanctions (WWF, 2016a), a measure that could be read as calling into question the country's international legitimacy.

These pressures were echoed throughout the Vietnamese media coverage of global IWT events, including the 2016 Hanoi IWT conference. This media coverage framed IWT as a serious crime (Massé et al., 2020) and drew attention to Vietnam's failure to commit to meaningful, on the ground action (i.e. using framings such as Serious Crime, Compliance and Lack of Enforcement). Such international events can serve as opportunities for frame establishment (Gitlin, 1980; Krebs, 2015) as the media are particularly drawn to event-driven narratives (Anderson, 1997). Media efforts at the international and civil society level may have educated the public and enforcement agencies and facilitated Vietnam's implementation of international treaty CITES (Tallberg, 2002: 617; Mitchell, 2003: 450–451; Spector & Korula, 1993: 381). As the holding of the conference was described as symbolic of Vietnam's commitment and determination to fight IWT, Vietnamese public officials may have also used local media coverage to employ the Serious Crime and Compliance framing to reclaim legitimacy (Solesbury, 1976).

Framing evolves over time and often requires preparatory background framing (Lakoff, 2010). In the lead-up to the 2016 events, NGOs contributed to this process by promoting Responsibility and Lip Service frames, which depicted rhinos as victims of brutality (Smith, 2022) rather than as property (Sollund, 2022). Subsequently, UNTOC guidelines issued through UNODC appear to have shaped the Serious Crime frame, while aspects of the Compliance frame reflected CITES' influence. Vietnamese digital newspapers' emphasis on the country's determination to strengthen legal frameworks and international cooperation may have drawn legitimacy from these earlier frames (Solesbury, 1976). This is underscored by UNODC and CITES being cited as elite sources in Vietnamese media, consistent with Gitlin's (1980) observation that media framing reflects dominant institutional forces. In addition, media play an important role in educating both the public and policymakers, thereby facilitating the implementation of international agreements (Spector & Korula, 1993).

Given that all media in Vietnam is state-owned, the inclusion of quotes from international and environmental organisations suggests the potential for supranational actors to influence local policy and foster a growing green public sphere (Papastergiadis, 2000; Yang & Calhoun, 2013). The media deliberations on the rhino horn trade in Vietnam illustrate how rising global media and public attention can educate citizens, law enforcement agencies, and government officials, 'especially in cases where the laws are hard to understand because the enforcement agencies do not enforce them' (Goyes & Sollund, 2016: 90). In this way, media attention can contribute to shifting national practices aimed at reducing environmental harm (Mitchell, 2003).

The criminality framing of the rhino horn trade in the Vietnamese media emphasises the certainty and severity of punishment, central themes in deterrence literature (Braga & Weisburd, 2012: 1). This framing is reinforced through reports of multiple seizures, tougher sentences, effective policing, coordinated enforcement nationally and internationally, and coverage of the modus operandi of criminal syndicates. Such a criminal justice approach aligns with aspects of green criminology by potentially deterring those who contribute to harm against rhinos. Yet shifting responsibility for addressing the rhino horn trade from conservation actors to enforcement and security agencies raises concerns (Massé et al., 2020). Criminality-focused responses may be limited in effectiveness (Wellsmith, 2011: 134), given persistent challenges such as under-resourcing, marginalisation within the broader criminal justice system, the hidden scale of the trade, corruption, and a general failure to treat wildlife crime as serious. These factors collectively undermine deterrence, and even where barriers are addressed, prosecutions and harsher penalties alone have little impact on reducing IWT (Wellsmith, 2011). Alternative strategies, such as promoting sustainable livelihoods, creating economic incentives for species protection (Sollund, 2022: 1030), and reducing consumer demand, may provide more effective long-term solutions, particularly when integrated with enforcement (Burgess, 2016). Supporting this view, Sollund (2022: 1030) argues from a green criminological perspective that wildlife trade crimes should nonetheless be prosecuted and punished proportionately to the harms caused. While enforcement remains the dominant response, the news reporting indicates that Vietnam must demonstrate more concrete action, particularly through international cooperation.

6 Conclusion

News framing analysis helps reveal what is at stake in transnational environmental issues. Sustained framing efforts by international environmental NGOs and elites, particularly through the Lip Service, Serious Crime, Compliance, and Lack of Enforcement frames, helped shape Vietnamese media narratives and applied pressure on the state to take stronger action against the rhino horn trade. Symbolic events, such as Vietnam's destruction of ivory and rhino horn, amplified these frames, while endorsement from global institutions such as CITES and UNODC lent them legitimacy. Together, these dynamics contributed to legal reforms, harsher sentencing, and heightened enforcement. Through the adoption of criminality frames, international actors influenced local media, prompting shifts in state response, while Vietnam reciprocated with its own framing to signal a bid to reclaim legitimacy.

International pressure from stakeholders and NGOs appears to have pushed Vietnam to treat IWT more seriously, with prosecutions now occurring where previously there were none. The certainty of arrest and punishment is increasing; and courts are imposing sentences that approach proportionality to the harms inflicted on rhinos. Although Vietnam's justice system remains human-centred, these developments offer a form of justice for rhinos, primarily through deterrence. Deterrence logic is visible in Vietnam's adoption of CITES commitments and its classification of rhino horn trade as a serious crime. Yet the persistence of trafficking casts doubts on the full deterrent effect of these measures. True justice for rhinos remains elusive, as poaching in Africa continues to threaten their survival. Moving beyond criminality frames to those recognising rhinos as sentient beings with inherent rights may advance justice more meaningfully. Until then, reducing their harm and suffering may be best served by a mixed approach, combining conservation, deterrence, demand reduction and sustainability methods, without allowing enforcement and security agendas to consume a disproportionate share of resources.

Acknowledgements I would like to acknowledge my supervisors Dr Kasun Ubayasiri, Dr Li Eriksson and Professor Mark Pearson at Griffith University for their support and guidance in gaining my PhD.

References

Anderson, A. (1997). *Media, Culture and the Environment*. UCL Press.

Braga, A., & Weisburd, D. (2012). The Effects of "Pulling Levers" Focused Deterrence Strategies on Crime. *Campbell Systematic Reviews, 6*. https://doi.org/10.4073/csr.2012.6

Brisman, A. (2018). Representing the "Invisible Crime" of Climate Change in an Age of Post-Truth. *Theoretical Criminology, 22*(3), 468–491. https://doi.org/10.1177/1362480618787168

Brisman, A., & South, N. (2014). *Green Cultural Criminology: Constructions of Environmental Harm, Consumerism, and Resistance to Ecocide*. Routledge.

Burgess, G. (2016). Powers of Persuasion? *TRAFFIC Bulletin, 28*(2), 65.

Dirikx, A., & Gelders, D. (2009). Global Warming Through the Same Lens: An Exploratory Framing Study in Dutch and French Newspapers. In T. Boyce, J. Lewis, & P. Lang (Eds.), *Media and Climate Change* (pp. 200–210). Oxford.

Drury, R. (2011). Hungry for Success: Urban Consumer Demand for Wild Animal Products in Vietnam. *Conservation and Society, 9*(3), 247–257. https://doi.org/10.4103/0972-4923.86995

Entman, R. M. (1993). Framing: Toward Clarification of a Fractured Paradigm. *Journal of Communication, 43*(4), 51–58. https://doi.org/10.1111/j.1460-2466.1993.tb01304.x

Education for Nature Vietnam Environmental Investigation Agency (EIA). (2013). Vietnam's Illegal Rhino Horn Trade: Undermining the Effectiveness of CITES. Retrieved May 15, 2025, from https://eia-international.org/wp-content/uploads/Species-Rhino-CITES-COP-Red1.pdf

Education for Nature Vietnam. Retrieved June 26, 2026, from https://www.facebook.com/EducationforNatureVietnam

Environmental Investigation Agency (EIA). (2021). Vietnam's Footprint in Africa: An Analysis of the Role of Vietnamese Criminal Groups in Wildlife Trafficking. Retrieved May 15, 2025, from https://eia-international.org/wp-content/uploads/EIA-Report-Vietnams-Footprint-in-Africa-Spreads-2021-11.pdf

Fraser, N. (2007). Transnationalizing the Public Sphere: On the Legitimacy and Efficacy of Public Opinion in a Post-Westphalian World. *Theory, Culture & Society, 24*(4), 7–30. https://doi.org/10.1177/0263276407080090

Gamson, W. A., & Modigliani, A. (1989). Media Discourse and Public Opinion on Nuclear Power: A Constructionist Approach. *American Journal of Sociology, 95*(1), 1–37. https://doi.org/10.1086/229213

Garstecki, T. (2006). Implementation of Article 16, Council Regulation (EC) No. 338/97 in the 25 Member States of the European Union. A TRAFFIC Report for the European Commission. TRAFFIC Europe, Retrieved May 5, 2025, from https://policycommons.net/artifacts/1935445/implementation-of-article-16-council-regulation-ec-no/2687215/

Gitlin, T. (1980). *The Whole World is Watching: Mass Media in the Making and Unmaking of the New Left*. University of California Press.

Goyes, D. R., & Sollund, R. (2016). Contesting and Contextualising CITES: Wildlife Trafficking in Colombia and Brazil. *International Journal for Crime, Justice and Social Democracy, 5*(4), 87–102. https://doi.org/10.5204/ijcjsd.v5i4.331

Hanoi Conference on Illegal Wildlife Trade. (2016). Hanoi Statement on Illegal Wildlife Trade. https://www.traffic.org/site/assets/files/2808/hanoi-statement-on-illegal-wildlife-trade.pdf

Hansen, A. (2011). Communication, Media and Environment: Towards Reconnecting Research on the Production, Content and Social Implications of Environmental Communication. *International Communication Gazette, 73*(1–2), 7–25. https://doi.org/10.1177/1748048510386739

High, M. D. (2023). The Perils and Potential Benefits of Machine Translation in Transnational Higher Education. In G. R. Morris & L. Li (Eds.), *Handbook of Research on Developments and Future Trends in Transnational Higher Education* (pp. 115–136). IGI Global.

Hertog, J. K., & McLeod, D. M. (2001). A Multiperspectival Approach to Framing Analysis: A Field Guide. In S. D. Reese, O. H. Gandy Jr., & A. E. Grant (Eds.), *Framing Public Life: Perspectives on Media and Our Understanding of the Social World* (pp. 139–162). Lawrence Erlbaum.

Huong, G. (2020, July 14). Surprise from Wildlife Crime Report 2020. *Lao Dong*.

Knupfer, C. B., & Entman, R. M. (2018). Framing Conflicts in Digital and Transnational Media Environments. *Media, War & Conflict, 11*(4), 476–488. https://doi.org/10.1177/1750635218796381

Krebs, R. R. (2015). Tell Me a Story: FDR, Narrative, and the Making of the Second World War. *Security Studies, 24*(1), 131–170. https://doi.org/10.1080/09636412.2015.1001215

Lang, Q. (2019, October 27). Why do Vietnamese People have to go to Jail Abroad for Ivory Bracelets? *Lao Dong.*

Lakoff, G. (2010). Why it Matters How We Frame the Environment. *Environmental Communication, 4*(1), 70–81. https://doi.org/10.1080/17524030903529749

Le Quang, V. (2019, December 16). Young Female Teacher Won the Contest "The Person Changing Perception of Rhino". *Lao Dong.*

Lester, L. (2014). Transnational Publics and Environmental Conflict in the Asian Century. *Media International Australia, 150*(1), 167–178. https://doi.org/10.1177/1329878X1415000128

Lever, G. (2017). UK–Vietnam Work Together to Tackle Illegal Wildlife Trade. Retrieved May 15, 2025, from https://www.gov.uk/government/speeches/work-together-to-tackle-illegal-wildlife-trade

Loughran, T. A., Paternoster, R., & Weis, D. B. (2016). Deterrence. In A. R. Piquero (Ed.), *The Handbook of Criminological Theory* (pp. 50–74). Wiley.

Lynch, M. J., Long, M. A., & Stretesky, P. B. (2021). Green Criminology. In B. Schaefer Caniglia, A. Jorgenson, S. A. Malin, L. Peek, D. N. Pellow, & X. Huang (Eds.), *Handbook of Environmental Sociology* (pp. 355–379). Springer.

Massé, F., & Margulies, J. D. (2020). The Geopolitical Ecology of Conservation: The Emergence of Illegal Wildlife Trade as National Security Interest and the Re-Shaping of US Foreign Conservation Assistance. *World Development, 132,* 104958. https://www.sciencedirect.com/science/article/pii/S0305750X2030084X

Massé, F., Dickinson, H., Margulies, J., Joanny, L., Lappe-Osthege, T., & Duffy, R (2020). Conservation and Crime Convergence? Situating the 2018 London Illegal Wildlife Trade Conference. *Journal of Political Ecology, 27*(1), 23–42. https://doi.org/10.2458/v27i1.23543

Milliken, T., & Shaw, J. (2012). The South Africa–Viet Nam Rhino Horn Trade Nexus: A Deadly Combination of Institutional Lapses, Corrupt Wildlife Industry Professionals and Asian Crime Syndicates. TRAFFIC. Retrieved May 15, 2025, from https://www.traffic.org/site/assets/files/2662/south_africa_vietnam_rhino_horn_nexus.pdf

Mitchell, R. B. (2003). International Environmental Agreements: A Survey of their Features, Formation, and Effects. *Annual Review of Environment and Resources, 28*(1), 429–461. https://doi.org/10.1146/annurev.energy.28.050302.105603

Nagin, D. S. (2013). Deterrence: A Review of the Evidence by a Criminologist for Economists. *Annual Review of Economics, 5*(1), 83–105. https://doi.org/10.1146/annurev-economics-072412-131310

Nagin, D. S. (2018). Deterrent Effects of the Certainty and Severity of Punishment. In D. S. Nagin, F. T. Cullen, & C. L. Jonson (Eds.), *Deterrence, Choice, and Crime* (Vol. 23, pp. 157–185). Routledge.

Nagin, D. S., & Pogarsky, G. (2001). Integrating Celerity, Impulsivity, and Extralegal Sanction Threats into a Model of General Deterrence: Theory and Evidence. *Criminology, 39*(4), 865–892. https://doi.org/10.1111/j.1745-9125.2001.tb00943.x

Nagin, D. S., & Pogarsky, G. (2003). An Experimental Investigation of Deterrence: Cheating, Self-Serving Bias, and Impulsivity. *Criminology, 41*(1), 167–194. https://doi.org/10.1111/j.1745-9125.2003.tb00985.x

Ngoc, Y. (2018, July 19). 30 Months of Action Against the Ivory, Rhino Horn Trade. *Cong An Nhan Dan.*

Nhu, A. (2020, 29 December). Sentenced for Illegally Transporting 11 Rhino Horns. *Lao Dong.*

Nisbet, M. C. (2009). Communicating Climate Change: Why Frames Matter for Public Engagement. *Environment: Science and Policy for Sustainable Development*, *51*(2), 12–23. https://doi.org/10.3200/ENVT.51.2.12-23

Pan, Z., & Kosicki, G. M. (2001). Framing as a Strategic Action in Public Deliberation. In S. D. Reese, O. H. Gandy, Jr., & A. E. Grant (Eds.), *Framing Public Life: Perspectives on Media and Our Understanding of the Social World* (pp. 35–66). Lawrence Erlbaum.

Papastergiadis, N. (2000). *The Turbulence of Migration.* Polity.

Pham, T. T., Di Gregorio, M., & Brockhaus, M. (2017). REDD+ Politics in the Media: A Case Study from Vietnam. *International Forestry Review, 19*(1), 69–80. https://doi.org/10.1505/146554817822407466

Save the Rhino. (n.d.). *Poaching Statistics.* Retrieved May 15, 2025, from https://www.savetherhino.org/rhino-info/poaching-stats/

Scanlon, J. E. (2016). Remarks on the Destruction of Confiscated Elephant Ivory and Rhino Horn in Hanoi, Viet Nam. *CITES*. Retrieved May 15, 2025, from https://cites.org/eng/CITES_SG_remarks_destruction_ivory_rhino_horn_vietnam_12112016

Semetko, H. A., & Valkenburg, P. M. (2000). Framing European Politics: A Content Analysis of Press and Television News. *Journal of Communication, 50*(2), 93–109. https://doi.org/10.1111/j.1460-2466.2000.tb02843.x

Smith, M. S. (2022). You Better Believe It: Rhino Horns are a Bloody Risky Business. In E. Freedman, S. S. Hiles, & D. B. Sachsman (Eds.), *Communicating Endangered Species: Extinction, News and Public Policy* (pp. 116–134). Routledge.

Smith, M. S. (2024). *The Rhino Horn Trade, Transnational Organised Crime and Vietnamese State-Run Media: The News Framing of Enforcement of Laws Against the Illegal Trade in Rhinoceros Horn in the Vietnamese Digital Newspapers* (Doctoral Thesis). Griffith University. https://doi.org/10.25904/1912/5609

Smith, M. S. (2025). News Framing of Environmental Crime. In R. White (Ed.), *Elgar Encyclopedia of Environmental Crime* (pp. 63–66). Edward Elgar.

Solesbury, W. (1976). The Environmental Agenda: An Illustration of How Situations May Become Political Issues and Issues May Demand Responses From Government: Or How They May Not. *Public Administration, 54*(4), 379–397. https://doi.org/10.1111/j.1467-9299.1976.tb00256.x

Sollund, R. (2022). Wildlife Trade and Law Enforcement: A Proposal for a Remodeling of CITES Incorporating Species Justice, Ecojustice, and Environmental Justice. *International Journal of Offender Therapy and Comparative Criminology, 66*(9), 1017–1035. https://doi.org/10.1177/0306624X221099492

Spector, B. I., & Korula, A. R. (1993). Problems of Ratifying International Environmental Agreements: Overcoming Initial Obstacles in the Post-Agreement Negotiation Process. *Global Environmental Change, 3*(4), 369–381. https://doi.org/10.1016/0959-3780(93)90025-G

Tallberg, J. (2002). Paths to Compliance: Enforcement, Management, and the European Union. *International Organization, 56*(3), 609–643. https://doi.org/10.1162/002081802760199908

Thien, M. (2018, March 22). Proposal to Appeal First Instance Verdict for "Oligarchs" Smuggling Rhino Horn. *Lao Dong*.

Tuong, P. (2018, January 25). The Tycoon in the Largest Wildlife Trade in Thai History Fell into the Net. *Cong An Nhan Dan*.

Turner, A. M., Bergman, M., Brownstein, M., Cole, K., & Kirchhoff, K. (2014). A Comparison of Human and Machine Translation of Health Promotion Materials for Public Health Practice: Time, Costs, and Quality. *Journal of Public Health Management and Practice, 20*(5), 523–529. https://doi.org/10.1097/PHH.0b013e3182a95c87

United Nations Office on Drugs and Crime (UNODC). (2004). United Nations Convention Against Transnational Organized Crime and the Protocols Thereto. Retrieved May 5, 2025, from https://www.unodc.org/documents/treaties/UNTOC/Publications/TOC%20Convention/TOCebook-e.pdf

Van Gorp, B. (2007). The Constructionist Approach to Framing: Bringing Culture Back in. *Journal of Communication, 57*(1), 60–78. https://doi.org/10.1111/j.1460-2466.2006.00329_3.x

Van Song, N. (2008). Wildlife Trading in Vietnam: Situation, Causes, and Solutions. *The Journal of Environment & Development, 17*(2), 145–165. https://doi.org/10.1177/1070496508316220

Voung, T. (2016, June 1). Fine Up to 1.5 Billion VND for Crimes Related to Wildlife Trade. *Lao Dong*.

Vy, K. (2016, September 17). World Calls on Vietnam to Reduce Consumption of Rhino Horn. *Cong An Nhan Dan*.

Wellsmith, M. (2011). Wildlife Crime: The Problems of Enforcement. *European Journal on Criminal Policy and Research, 17*, 125–148. https://doi.org/10.1007/s10610-011-9140-4

Wichgers, L., Jacobs, L., & van Spanje, J. (2021). The Battle of Frame Building: The Reciprocal Relationship Between Journalists and Frame Sponsors. *The International Journal of Press/Politics, 26*(3), 674–698. https://doi.org/10.1177/1940161220942760

World Wide Fund for Nature. (2016a). CITES Increases Pressure on Vietnam Over Illegal Rhino Horn Trade. Retrieved May 15, 2025, from https://wwf.panda.org/wwf_news/?279930/CITES-increases-pressure-on-Vietnam-over-illegal-rhino-horn-trade

World Wide Fund for Nature (2016b). WWF Reaction to the Hanoi Statement on Wildlife Crime. Retrieved May 15, 2025, from https://www.wwf.sg/wwf-reaction-to-the-hanoi-statement-on-wildlife-crime/

World Wide Fund for Nature. (2019). More Than Words: Are Commitments to Tackle Illegal Wildlife Trade Being Met? Retrieved May 15, 2025, from https://www.wwf.org.uk/sites/default/files/2019-08/WWF_IWT_Report_v8.pdf

Wyatt, T. (2014). Assessing the Partner and Media Engagement with the International Consortium on Combating Wildlife Crime. In R. White, M. Kluin, & T. Spapens (Eds.), *Environmental Crime and Its Victims: Perspectives Within Green Criminology* (pp. 277–291). Ashgate Publishing.

Yang, G., & Calhoun, C. (2013). Media, Civil Society and the Rise of the Green Public Sphere in China. In L. Lester, & B. Hutchins (Eds.), *Environmental Conflict and the Media* (pp. 273–290). Peter Lang.

11

Visions of 'Justice' in the Campaigns to Criminalise Ecocide

Rachel Killean

1 Introduction

In this chapter, I examine the concept of 'justice' in the context of the contemporary campaign to introduce 'ecocide' as a 5th crime against peace at the International Criminal Court ('ICC'), and the related campaigns to criminalise ecocide in a range of domestic jurisdictions. I consider two questions. First, what visions of 'justice' are campaigners pursuing when they advocate for ecocide's criminalisation? Second, is criminalisation equipped to deliver these visions of justice in practice? To contextualise these questions, I briefly review the history of these movements, charting the emergence of ecocide as a concept in the 1970s, through to the Republic of Vanuatu's 2024 proposal to amend the ICC's Rome Statute to include the crime of ecocide (Part 2). I also outline the approaches adopted amongst domestic states, noting the diversity in approach, the changing conceptualisation of ecocide over time, the

R. Killean (✉)
University of Sydney, Sydney, NSW, Australia
e-mail: rachel.killean@sydney.edu.au

R. Killean et al. (eds.), *Conceptualising Justice*, Palgrave Studies in Green Criminology, https://doi.org/10.1007/978-3-032-19791-7_11

growing momentum behind the Stop Ecocide Foundation's Independent Expert Panels' 2021 definition, and European regional developments. To explore the visions of justice within these campaigns, I analyse campaigners' media outputs and other public-facing documents (Part 3). I identify four dominant visions of justice that are offered as justification for the criminalisation of ecocide: preventative justice (deterrence); retributive justice (punishment); transformative justice (societal transformation); and expressive justice (symbolic condemnation). As I explore, these framings of justice find resonance with green criminology literature that advocates for greater use of criminal law as a means of pursuing justice for environmental harms.

In Part 4, I draw on critical green criminology, criminology of international crime, and legal literature to reflect on the second question: whether criminalisation is equipped to deliver these visions of justice in practice. I highlight two key critiques that emerge in the literature: that evidence of criminal law's deterrent potential is inconclusive, and that there is a problematic distinction between the reality of ecocide and the formulation of ecocide as a crime. As I explore, these critiques substantially challenge the justice narratives put forward by the ecocide campaigns. In Part 5, I offer some concluding reflections on the value and limitations of criminalisation as a means of responding to the destruction of the environment. Drawing on White's (2024) notion of 'innovative justice' and McEvoy's (2007) notion of 'legal humility', I argue there is a need for more nuanced engagement with what ecocide's criminalisation can achieve. My goal in highlighting these limitations is not to undermine the important efforts of activists in this space. Rather, my goal is to acknowledge that ecocide's public appeal might exceed its potential impact, and to encourage the pursuit of joined-up legal and non-legal responses to ecological crisis. Far from being 'the only way' to stop climate disaster (as argued by prominent ecocide campaigners, Mehta & Jackson, 2021), criminalisation needs to be understood as one small piece of broader efforts to prevent the destruction of our planet.

2 Charting the Ecocide Campaigns

Since emerging in the 1970s, ecocide has transitioned from a concept rooted in war-related environmental destruction to one linked to the global climate and ecological crises. In this brief review of its history I explore the key developments, shifting conceptualisations, and diverse legal approaches that have characterised the campaign over the last fifty years.

2.1 The Birth of Ecocide

The concept of 'ecocide' first emerged as a descriptor for the United States' use of defoliant herbicides in the Vietnam War (Zierler, 2011). The violent impacts of herbicidal use caused international shockwaves, leading to calls for greater regulation around the use of environmental warfare (Falk, 1973; Fried, 1973). Although the term 'ecocide' itself does not appear in the legal documents that followed these calls for regulation, explicit environmental protections made their way into international humanitarian law through the Additional Protocol I to the 1949 Geneva Conventions, and the Convention on the Prohibition of Military or Any Other Hostile Use of Environmental Modification Techniques (ENMOD). Debates regarding the appropriate response to severe environmental crime continued throughout the 1970s, 80s and 90s (Ruhashyankiko, 1978; Tomuschat, 1996; Whitaker, 1985). During this time period, some advocated for an international environmental crime against humanity or autonomous environmental crime, but this was met with resistance from states who were reluctant to see criminalisation extend beyond wartime contexts (Tomuschat, 1996). By the time international criminal law came to be codified in the ICC's Statute in 1998, the debate had largely settled, with the only explicit prohibition on environmental destruction found in the context of war crimes (ICC, 1998: Art. 8(2)(b)(iv)). At a domestic level, a small number of states adopted a broader approach, prohibiting environmental crimes against humanity (Vietnam, 1985) and ecocide as a crime against peace (all former USSR states: the Russian Federation, 1996; Kazakhstan and the

Kyrgyz Republic, 1997; Tajikistan, 1998; Belarus and Georgia, 1999; Ukraine, 2001; Moldova, 2002; and Armenia, 2003. See Killean, 2025 for an analysis of this trend).

2.2 Contemporary International Ecocide Campaigns

The campaign to criminalise ecocide quietened in the late 1990s and early 2000s (although see Drumbl, 1998; Gray, 1996) but rose to prominence again alongside increasingly urgent concerns about climate change. Polly Higgins is often attributed with kick-starting contemporary ecocide debates. Although unsuccessful in advocating for the International Law Commission to adopt a new crime of ecocide, Higgins' work sparked new grassroots movements (End Ecocide in Europe, 2012; End Ecocide on Earth, 2016), drew attention to ecocide's history (Gauger et al., 2012), and, perhaps most influentially, led to the creation of Stop Ecocide, now Stop Ecocide International ('SEI'). SEI's impacts have been widespread, including supporting Small Island States in participating at the annual Assembly of ICC State Parties, thereby enabling them to advocate for ecocide's international criminalisation (Mehta, 2019). At a grassroots level, SEI has established regional and domestic branches and facilitated support for local activists across a range of contexts (Schnyder, 2023).

In 2021, the charitable Foundation of SEI convened an Independent Expert Panel ('the Panel') of international criminal law and environmental law experts to develop a new definition of ecocide. The resulting definition, '*unlawful or wanton acts committed with knowledge that there is a substantial likelihood of severe and either widespread or long-term damage to the environment being caused by those acts*', draws on existing international criminal legal language, while extending the prospect of environmental criminality to times of peace (Hansen, 2025). Notwithstanding some critiques (e.g. Minkova, 2023), the recognised expertise of the Panel, the widespread activism of SEI, and the advocacy efforts of the Panel's members, have contributed to this definition achieving an almost hegemonic status in contemporary debates. It has since been taken up by Vanuatu, which in September 2024 submitted a formal proposal for an amendment to the ICC's Statute, following their initial proposal that such a crime be considered at the ICC's Assembly of State Parties in 2019.

Despite this growing momentum, substantial barriers to ecocide's inclusion in the ICC's Statute remain. An amendment requires at least a two third majority of the ICC's States Parties, with a final text of the proposed amendment requiring discussion and State Party agreement. While a final discussion could take place at a 'Crime Review Conference', this process has rarely been used. Alternatively, a 'Working Group on Amendments' could be tasked with evaluating a proposed amendment. This can be a long process; it took two decades for the crime of aggression to be adopted despite early agreement regarding its definition (Chin, 2023). Any proposal will face intense scrutiny, with unofficial and official channels of discussion creating space for political and corporate interests to influence the process (Killean & Short, 2025).

2.3 Domestic Developments

Perhaps in response to these barriers, a notable shift has occurred in SEI's advocacy since 2021, with increased emphasis placed on regional and domestic developments. The Panel's definition has certainly played a role in this. A wave of ecocide bills followed on the heels of the Panel's definition (e.g. Spain (Catalunya), Mexico, The Netherlands, Italy, Scotland and the UK, Brazil) with varying levels of success at the time of writing (see also Killean & Short, 2023). In Europe, the 2024 EU 'Directive on the Protection of the Environment through Criminal Law' and 2025 Council of Europe's *Convention on the Protection of the Environment through Criminal Law* acknowledge the growing momentum behind the ecocide campaign, and, more substantively, require States to criminalise conduct comparable to ecocide (Bertram, 2025). Belgium has already done so, approving an ecocide crime in 2024 (Chini, 2024). The crime is somewhat more narrowly worded than the Panel's definition, requiring illegality and serious, widespread and long-term damage to the environment. It is placed immediately after 'Grave Violations of International

Humanitarian Law' but applies 'whether it is committed in a time of peace or in a time of war' (Bertram, 2024).

There are also examples of domestic systems adopting their own approaches to severe environmental criminality. For example, in 2021 France introduced a 'Climate and Resilience Law', which introduced a qualified ecocide offence to the French Environmental Code. Although purportedly influenced by prominent French ecocide advocates as well as SEI (Artuad et al., 2023) the resulting legislation is somewhat different from SEI's approach—introducing qualified offences of pollution and abandoning waste rather than an autonomous offence.

2.4 A Contested and Contingent Campaign

This brief summary highlights two points. First, that while the Panel's definition has furthered greater cohesion within the ecocide campaign, the scope and definition of ecocide remain contested, varied, and contingent on legal and political contexts. Second, the context has changed since ecocide first emerged as a potential war crime. This is evident in the debates that have led to successful domestic criminalisation: in the French case, ecocide emerged from citizens' concerns over how to stem the harms of the climate crisis, while in Belgium, the debate centred on the biodiversity and climate crisis. As explored below, this changing context has directly shaped the visions of justice pursued by campaigners.

3 Visions of Justice

The following analysis is based on public-facing documents connected to campaigns to criminalise ecocide in different fora, gathered between March and May 2025. To explore the visions of justice that inform the campaign to criminalise ecocide at the ICC, I focused on the website of SEI, the largest international advocacy group pursuing ecocide's criminalisation. SEI's website acts as a repository for not only its own advocacy tools and opinions, but also guest blogs republished from other sources, and links to media articles published elsewhere from January 2024 onwards. I downloaded official SEI documents, such as an Executive Summary, a Youth Toolkit, and a Q&A section, documents associated with the Panel's 2021 definition, all available guest blogs, and all media links that were not paywalled.

I then sourced documents from states that have pursued criminalisation in the last five years: Belgium, Brazil, France, Mexico, The Netherlands, Scotland, Spain and the UK. I included the Explanatory Statement to the 2024 EU Environmental Crime Directive, alongside submissions from the European Parliament's Committees on Civil Liberties, Development, Petitions, and the Environment, and documents associated with the Council of Europe's 2025 Convention. In total, I analysed 45 documents. I make no claims to having conducted an exhaustive search or that this is the definitive account of the justice claims of ecocide's many proponents. Yet, even this limited analysis offers some insights into the conceptualisations of 'justice' that emerge across the various campaigns.

To code, I drew from green criminology and broader criminological literature to identify the justice 'goals' used to justify environmental harm's criminalisation. I looked for specific references to justice, but also words and descriptions of concepts associated with the aspirations of criminal justice. These included, for example, preventative justice (deterrence) (see e.g. Ashworth & Zedner, 2014), retributive justice (punishment) (see e.g. Moore, 2010), restorative justice (restoration) (see e.g. Braithwaite, 2002), reparative justice (repair, reparation) (see e.g. Daly & Proietti-Scifoni, 2012), expressive justice (symbolic condemnation, reflection of social values) (see e.g. Duff, 2001), and transformative justice (societal transformation) (see e.g. Corda, 2020). I remained open to additional concepts and themes emerging. Four dominant, interlinked visions of justice emerged from my coding: preventative, retributive, transformative and expressive justice. These are discussed in turn.

3.1 Preventative Justice

The most frequent vision of justice was preventative: i.e. the idea that criminalising ecocide would prevent future environmental destruction by deterring potential perpetrators. I coded references to deterrence more than twice as many times as any other justice framing (96 distinct references across 29 sources, compared to 42 distinct references to retributive justice across 16 sources, 38 to transformative justice across 17 sources, and 37 references to expressive justice across 15 sources). Many advocates of deterrence emphasised the necessity of ecocide's criminalisation given the ecological crises facing the planet. Criminalisation was framed as a valuable, self-evidently necessary tool for preventing future harm to the planet. For example:

> It is now necessary to take this step in Belgian criminal law in order to finally recognise the supreme interest of preserving the planet. (Chamber of Representatives of Belgium, 2021)
>
> The proposed Ecocide (Prevention) (Scotland) Bill seeks to deter mass environmental damage and destruction taking place in Scotland. It aims to protect the environment in Scotland covering all natural resources - air, water, soil, wild fauna and flora, including habitats. (Lennon, 2023)

The emphasis on deterrence reflects dominant approaches to criminal policy: deterrence is one of the most frequently mentioned reasons for introducing criminal sanctions (Orlando & Bergin, 2017: 8). It also reflects a common theme in the green criminology literature: that the 'widespread implications of environmental and social harm' require stronger enforcement tools (McKie et al., 2017: 42; see also South, 2017). Green criminologists have repeatedly drawn attention to the limitations of existing environmental laws and regulations in the absence of enforcement practices (Lynch et al., 2014; White, 2011). Similarly, I coded frequent explicit mentions of the failure of existing regulations to prevent the current ecological crises. For example:

> Only the recognition of ecocide as an international crime will prevent damage to the climate… [it will] strengthen existing legislation, including multilateral environmental agreements. (Jowa, 2024, citing Princess Esmerelda of Belgium)
>
> A standalone crime of ecocide is needed urgently… in recognition that international agreements to protect the environment that are without

> a legally enforceable deterrent are not delivering the urgent protections needed. (End Ecocide on Earth, 2024)

Another feature of the preventative justice discourse was the perceived potential for ecocide to effectively deter those in positions of power, such as corporate decision-makers and policymakers. For example:

> The crime of ecocide aims to curb the practice of acts planned and decided by people at the top of the chains of command in politics, the financial and corporate world, and the agroeconomy. (Brazil PSOL Party, 2023)

The emphasis on corporate and state perpetrators resonates with the crimes of the powerful strand of criminology, which for decades has sought to problematise the impunity with which such actors perpetrate social harms (Hillyard & Tombs, 2007; Schwendinger & Schwendinger, 1977; White, 2018). This approach is fundamental to the emergence of green criminology as a field, with criminologists arguing for a re-evaluation of who and what counts as 'criminal' when it comes to environmental harm, and for associated legal reform (Lynch et al., 2013; Westerhuis et al., 2013). Green criminology literature has argued that deterrence through criminal law may be particularly impactful on corporate actors, who are more likely to engage in cost–benefit analyses than perpetrators of other types of crime (e.g. interpersonal violence) and who may be unwilling to risk the social and reputational costs of prosecution (Du Rées, 2001). This was also reflected in my sources:

> Imagine sitting in the board room to discuss a fossil fuel project and the legal team says there's a risk of ecocide – it has a high deterrent value. (Koop, 2024, citing Independent Expert Panel member, Kate Mackintosh)

There has never been a crime quite like ecocide before, i.e. an *international* crime that puts people in positions of corporate power at risk of prosecution for business decisions that impact the environment. This means there may be a limit to the lessons that can be learned from existing deterrence literature. However, I will return to the question of deterrence in Part 4, arguing that the literature—at a minimum—suggests a need for nuanced claims in this regard.

3.2 Retributive Justice

In the literature, prevention is often linked to retribution and punishment as related justice goals, with criminal enforcement framed as a way to 'detect, deter and *punish*' (McKie et al., 2017: 41). This was also true in my analysis: retributive justice was the next most frequently mentioned vision of justice, i.e. the belief that the criminalisation of ecocide would enable perpetrators to be investigated, arrested, and appropriately punished. Harsher criminal sanctions were often linked to the notion of deterrence:

> Harsher sanctions for lawbreakers and stronger enforcement are urgently needed. The sanctions need to be high enough to have the desired deterrence effect. (European Parliament, Committee on the Environment, Public Health and Food Safety, 2022)
> Only by imposing significant criminal sanctions on those who damage our planet can we begin to prevent the catastrophe they are on their way to causing. (Ragnarsdóttir & Sharpston, 2023)

This belief in the links between severe sanctions and deterrence appears in many debates around how law should respond to environmental harms (Du Rées, 2001). It builds on the longstanding 'rational choice' criminological theory which posits that for criminal laws to deter, there must be a certain probability that offences will be investigated *and* lead to sufficiently severe sanctions (Akers, 1990). In green criminology, research has critiqued how environmental harms are 'trivialised', even when criminalised (Wellsmith, 2011; White, 2008), with some arguing that lenient sentencing undermines criminalisation's deterrent effect (Ogus & Abbott, 2002).

However, retributive justice was not always framed alongside deterrence. For some, the need to criminalise ecocide reflected the classic retributive logic of punishment as an imperative (Kant, 1838), meaning criminalisation was valued as a way of ending impunity and facilitating sanctions for their own sake:

> It's time to react and highlight these errors in our legislation so that they [perpetrators] are punished with the gravity of what this implies, let's not allow them to continue in impunity. (Barrón Perales, 2023)
> It is intuitively clear that significant environmental harms threaten not only environmental rights but also all other human rights, for present and future generations. There is a compelling argument that this consideration should be transposed into legal rules and enforcement. (Maddrick, 2024)

This argument was often visible in advocacy for an *international* crime of ecocide. Advocates highlighted the international, transboundary and intergenerational nature of the harm caused by ecocide, as well as the scale of impunity, seemingly suggesting a greater moral imperative to hold perpetrators accountable (see also Higgins et al., 2013). Such arguments resonate with the concept of Environmental Impunity, theorised by Cedillo and Le Clercq as a descriptor for the specific ways impunity functions regarding environmental harms. They argue that the impunity of powerful actors 'deepens irreversible damage perpetrated against ecosystems' and 'multiplies the vulnerability and suffering of victims' (Cedillo & Le Clercq, 2021: 116), an approach that resonates with a more victim-oriented understanding of retributive justice (retribution as satisfaction for victims).

The focus on ending impunity for environmental and human rights violations in the second quote above also resonates with broader 'anti-impunity' and 'human rights penalty' agendas that have shaped international human rights discourse since the 1990s. These interlinked agendas are premised on an expectation that states' '(most) coercive and carceral apparatus' should be weaponised to protect human rights (Mavronicola, 2024: 536), and that rights violations should therefore be characterised as crimes (Alston, 2023). These agendas' specific application to environmental harms can be situated within a frustration regarding political inaction regarding the climate and ecological crisis (Cedillo & Le Clercq, 2021: 116). A similar awareness emerged in my analysis, with several sources understanding ecocide's criminalisation as part of a shift towards political and societal transformation.

3.3 Transformative Justice

Possibly the most radical understanding of justice that emerged from my coding was that which understood ecocide's criminalisation as a tool for bringing about broader transformations in approaches to the natural world. Ecocide's criminalisation was linked to transforming global approaches to resource management laws and policies:

> There is a need for an extremely rapid transition - by 2050 at the latest - towards a circular, low-carbon economy that respects biodiversity...This requires a radical change in the ways we use energy, occupy and exploit space, consume resources and produce waste. It is in this global context that we must consider the role that criminal law, and particularly international criminal law, can play in this transformation of our vision of the world. (Born et al., 2022)
> Proceeding to agree a crime of ecocide could contribute to a change of consciousness, in support of a new direction, one that enhances the protection of the environment and supports a more collaborative and effective legal framework for our common future on a shared planet. (Independent Expert Panel, 2021)

It was also described as a method through which approaches to business could be fundamentally changed, with markets re-envisioned as a place where environmental stewardship was prioritised:

> By creating a powerful incentive to prioritize nature over short-term gain, the recognition of ecocide as a crime has the potential to transform the way in which business is conducted. Holding decision-makers accountable for the most extreme forms of environmental harm can spark a market shift toward favoring stewardship of our natural world. (Lennon & Schwartz, 2024)

And finally, as a pathway to influencing how members of the public relate to the natural world:

> It will also make citizens, not just companies, aware of potentially polluting actions and hopefully influence people's overconsumption. (Netherlands Consultation, 2024)
>
> The adoption of [an international] crime of ecocide... is the best tool to signal the importance of creating a shift in consciousness in the way we humans see our relationship to nature. (Stop Ecocide Italia, 2024)

This vision of justice can be situated in a broader shift towards 'progressive punitivism' (Aviram, 2020), in which criminal sanctions are understood as more than 'setting fundamental ground rules of societal behaviour' (Orlando & Bergin, 2017: 8). Instead, they are framed as a meaningful way of responding to a wide range of social ills and facilitating broader processes of transformation. For example, Corda (2020: 624) has explored how high-profile pollution cases in the US raised public sensibility, 'establishing the wrongfulness and blameworthiness of wilful and reckless environmental damages as a shared norm'. In the green criminology and ecocide literature, particular emphasis is placed on international criminal law's transformative power (Higgins, 2012). Framing ecocide as a tool of transformation takes criminalisation further away from the more 'traditional' goals of punishment and deterrence, evidencing a willingness to rethink 'what crime means' in a time of crisis and inaction (Floyd, 2015). However, this framing of justice remains bound to notions of deterrence, the implication being that corporate decision-makers, policymakers and the public will change their behaviour at least in part because of the declared risk of criminal sanction (Knox & Whyte, 2025). As will now be explored, this framing of justice is also connected to the symbolic value of criminal law, as captured by the concept of expressive justice.

3.4 Expressive Justice

The final framing of justice that emerged in my coding was expressive justice. Rather than seeing criminalisation as a force of change, proponents of expressive justice highlighted the importance of (international and domestic) criminal law reflecting existing contemporary social beliefs and values:

> Criminal law… protects the legal values considered essential to the social order. (Chamber of Representatives of Belgium, 2021)
> Criminal law has an important signalling function. Norms and criminalizations reflect prevailing views in society and sanction behaviour that conflicts with those views. Given current social views on the importance of climate, nature and the environment, the lack of criminal sanctions for serious environmental damage is no longer justified. (van Raan, 2023)

The expressive function of criminal law has long been cited in legal and criminological literature as justifying its use over administrative or civil sanctions. For example, in his work on the 'aims of criminal law', Hart noted that 'what distinguishes a criminal from a civil sanction is the judgment of community condemnation which accompanies and justifies its imposition' (Hart, 1958: 404). Even separated from any notions of deterrence or retribution, criminal law is seen as an appropriate tool for voicing societal disapproval of conduct (Glasgow, 2015), with international criminal law offering an articulation of the harms of gravest concern to the international community (Hodgson, 2024). From such a starting point, green criminologists and legal scholars have also noted that the criminalisation of environmental wrongdoing offers symbolic proof of the importance of environmental values to a community (Haines & Parker, 2018; White & Kramer, 2015). In the ecocide-specific literature, the symbolic role of international criminal law is highlighted as a strength: reflecting a societal shift in thinking. In a world of multiple ecological crises, it is deemed appropriate to reframe 'massive environmental wrongdoing not as a mere regulatory infraction, but rather as one of the gravest crimes warranting international concern' (Robinson, 2022: 318). Given that not all members of the international community necessarily feel this way, it seems there is an overlap between the expressive and transformative goals of ecocide's proponents: ecocide is envisioned as reflecting the beliefs of those who support it and as a tool to change the views of those who do not.

4 Delivering Visions of Justice

Having outlined four most prominent (and overlapping) visions of justice that emerged from my analysis of ecocide campaigns' public-facing documentation, this section turns to this chapter's second question: is ecocide's criminalisation equipped to deliver these justice outcomes in practice? To consider this question, I draw on critiques that have emerged in the critical strands of green criminological and legal scholarship regarding criminalisation in general and the criminalisation of ecocide in particular. I argue that these critiques strongly suggest the need for more nuance in claims regarding what a new crime can achieve.

The first critique is that framing ecocide's criminalisation as a tool for preventative and transformative justice risks overstating the impact of criminal sanctions on perpetration. Although deterrence is a frequently cited justification for criminal law, critical legal and criminological literature suggests a need for substantially more modest claims. For example, while some scholars have argued that international criminal law has had a deterrent effect on atrocity perpetration (Appel, 2018; Dancy, 2017; Jo & Simmons, 2016), evidence is mixed and inconclusive, with others questioning whether this is or could ever be the case (Hodgson, 2021; McAuliffe, 2012; Mégret, 2021). The research on the connections between retributive and preventative justice is also inconclusive, with some criminological research problematising claims that severe punishment increases deterrence, including in the context of environmental crime (Doob & Webster, 2003). In the context of domestic criminal law, the assumption that potential perpetrators will engage in cost–benefit analyses that can be influenced by the threat of criminal sanction has been substantially critiqued (Knight, 2020).

Proponents of ecocide have emphasised that a particularly strong deterrent effect will result from introducing an international crime that specifically targets corporate and political decision-makers. It is possible that this will prove to be the case. As noted above, ecocide is relatively unusual in combining international condemnation with the specific targeting of senior corporate actors. Yet, research into the prevention of crimes of the powerful has found little evidence of criminal law's deterrent effects (Sarre, 2024; Schell-Busey et al., 2016). In their critique of the ecocide campaign, Knox and Whyte (2025) highlight the various ways in which corporate actors are able to neutralise and offset accusations of criminality, suggesting that it is their money and power that will limit deterrence, rather than enhance it (see also Whyte, 2016). With the evidence regarding deterrence in environmental crime, crimes of the powerful, and international criminal law all relatively inconclusive, there is reason to be cautious about over-claiming ecocide's potential deterrent effect.

The second critique is that campaigners over-simplify ecocide perpetration, with implications for almost all the justice claims made above. In keeping with green criminology's willingness to look beyond 'crime' to consider social harms, many green criminologists view ecocide as a descriptor of a harmful phenomenon, rather than a legal category. This literature has emphasised the slow, structural, and systematic ways in which ecocide is perpetrated, with ecocidal acts and omissions understood as deeply intertwined with the capitalist 'treadmill of production' (e.g. Crook et al., 2018; Lynch et al., 2020). This reality stands in stark contrast to the goals of the ecocide campaigns outlined above, which emphasise 'unlawfulness' and acts which are characterised by their abnormality and direct links to spectacular harm, perpetrated with intent or reckless disregard for the harm caused.

This critical scholarship argues that criminal law's emphasis on individual perpetration is ill-suited to the 'slow violence' (Nixon, 2011) and complexity of environmental destruction (Knox & Whyte, 2025). As such, both preventative and retributive goals may be undermined by the difficulties in attributing blame to individual persons, while the 'expressive justice' envisioned may (unintentionally) narrow the behaviour being condemned. Furthermore, criminal law exists within, not separate to, a broader economic and political context. Currently, both are dominated by neoliberal capitalism and extractivism, both of which in turn facilitate and necessitate environmental destruction. Criminalising ecocide is therefore limited in its transformative potential in the absence of more fundamental shifts in political economic approaches (Cusato & Jones, 2023).

5 Towards Innovation and Humility

This chapter has outlined recent campaigns to criminalise ecocide in domestic and/or criminal law. It has examined the visions of justice that emerge in the campaigns' public-facing documentation, finding that the four most frequently mentioned forms of justice are preventative, retributive, transformative, and expressive. It has sought to contextualise those visions of justice in the green criminology literature, finding that all four visions of justice resonate with longer-standing debates about the role of criminal law in addressing environmental harms. Finally, it has turned to the more critical criminological and legal scholarship to consider whether ecocide's criminalisation is capable of delivering on those visions of justice. It found that some key critiques of criminal law were likely to offer significant challenges in doing so.

Despite the critiques offered in Part 4, I am not going to conclude by arguing that ecocide campaigners should abandon their pursuit of criminalisation. Although I share Knox and Whyte's (2025: 14) concerns that the ecocide campaign takes energy that might be spent on more transformative political projects, I also accept that ecocide seems to have a clear rhetorical force capable of mobilising a range of constituencies and political actors around the world. As such, it creates a discursive space to engage with environmental harm, which should not be dismissed as a benefit in its own right. Further, this chapter's modest review of the literature suggests that there are *uncertainties* and *limitations* regarding criminal law's ability to deliver on campaigners' visions of justice, which is not the same as saying no justification for criminalisation exists.

Yet, the institutional channels through which criminal law operates are narrow, contingent, and politically mediated. It seems clear that ecocide's public appeal as a label commensurate with the environmental crises may outweigh the practical impact of its criminalisation by a significant margin. My conclusion, therefore, draws from McEvoy's (2007) concept of 'legal humility'. Legal humility cautions against grand promises, and repositions criminal justice as one tool among many, to be used sparingly and alongside a wider range of non-legal processes. In the context of ecocide, legal humility necessitates greater transparency about the uncertainties and limitations of criminalisation. Such a step might be facilitated by engagement with Rob White's notion of 'innovative justice,' which reframes criminalisation not as the 'only way' (Mehta & Jackson, 2021) but as a modest contribution to a spectrum of interventions designed to protect the natural world, which might encompass civil and administrative enforcement, corporate governance and due-diligence regimes, reparative and community-led initiatives, Indigenous law and stewardship, and transnational environmental governance. Framed in this way, the promise of ecocide law presents not a silver bullet but a limited, carefully targeted contribution to a broader transformation in how polities value and protect the natural world.

References

Akers, R. L. (1990). Rational Choice, Deterrence, and Social Learning Theory in Criminology: The Path Not Taken. *The Journal of Criminal Law & Criminology, 81*(3), 653–676. https://doi.org/10.2307/1143850

Alston, P. (2023). Criminalizing Human Rights. *Journal of Human Rights Practice, 15*(3), 660–677. https://doi.org/10.1093/jhuman/huad059

Appel, B. J. (2018). In the Shadow of the International Criminal Court: Does the ICC Deter Human Rights Violations? *The Journal of Conflict Resolution, 62*(1), 3–28. https://doi.org/10.1177/0022002716639101

Artaud, M., de Froidefond, M., Gerbaux, J., Goux, H., Ribadeau Dumas, A., Samakh-Lesec, E., Toledo, M., & Venisse, N. (2023). L'Écocide en France: La Difficile Naissance Juridique d'une Notion Militante. *Les Controverses de Mines Paris.* https://controverses.minesparis.psl.eu/public/promo22/PC5Ecocide.pdf

Ashworth, A., & Zedner, L. (2014). *Preventive Justice.* Oxford University Press.

Aviram, H. (2020). Progressive Punitivism: Notes on the Use of Punitive Social Control to Advance Social Justice Ends. *Buffalo Law Review, 68*, 199–245.

Barrón Perales, K. (2023). *Initiative With a Draft Decree to Add Article 417 Bis to The Federal Penal Code, Regarding the Destruction of the Environment* (Ecocide).

Bertram, D. (2024, March 9). Ecocide à la Bruxelloise. *Verfassungsblog.* https://verfassungsblog.de/ecocide-a-la-bruxelloise/

Bertram, D. (2025, June 4). The First Ecocide Treaty? *EJIL: Talk!*, https://www.ejiltalk.org/the-first-ecocide-treaty/

Born, C., De Herdt J., Joëlle, R., & Vandermeersch, D. (2022). Towards the Inclusion of Ecocide as a Crime Under International Law in the New Penal Code. *Journal of the Courts, 6898*(18), 297–307.

Braithwaite, J. (2002). *Restorative Justice and Responsive Regulation.* Oxford University Press.

Brazil PSOL Party. (2023). *Bill No 2933/2023*, presented 5 June 2023.

Cedillo, C., & Le Clercq, J. A. (2021). Green Impunity: Measuring Ecojustice, Institutional Capacities and Policy Design as an Approach to Environmental Security. In A. Swain, J. Öjendal, & A. Jägerskog (Eds.), *Handbook of Security and the Environment* (pp. 115–131). Edward Elgar.

Chamber of Representatives of Belgium. (2021). *Proposed Law Aimed at Introducing the Crime of Ecocide into the Criminal Code,* 1 December 2021, Doc. 55 2356/001.

Chin, S. (2023). Lessons Learned from the Adoption of the Crime of Aggression: Ecocide to Charter its own Path. An International Crime of Ecocide: New Perspectives Symposium 2023. https://ecocidelaw.com/wp-content/uploads/2023/08/5-Chin-Lessons-learned-from-the-adoption-of-the-crime-of-aggression.pdf

Chini, M. (2024, February 23). Belgium Becomes First in EU to Recognise Ecocide as International Crime. *The Brussels Times.* https://www.brusselstimes.com/937229/belgium-becomes-first-in-eu-to-recognise-ecocide-as-international-crime-tbtb

Corda, A. (2020). The Transformational Function of the Criminal Law: In Search of Operational Boundaries. *New Criminal Law Review, 23*(4), 584–635. https://doi.org/10.1525/nclr.2020.23.4.584

Crook, M., Short, D., & South, N. (2018). Ecocide, Genocide, Capitalism and Colonialism: Consequences for Indigenous Peoples and Glocal Ecosystems Environments. *Theoretical Criminology, 22*(3), 298–317. https://doi.org/10.1177/1362480618787176

Cusato, E., & Jones, E. (2023). The "Imbroglio" of Ecocide: A Political Economic Analysis. *Leiden Journal of International Law, 37*(1), 1–20. https://doi.org/10.1017/S0922156523000468

Daly, K., & Proietti-Scifoni, G. (2012). Reparation and Restoration. In M. Tonry (Ed.), *The Oxford Handbook of Crime and Criminal Justice* (pp. 207–253). Oxford University Press.

Dancy, G. (2017). Searching for Deterrence at the International Criminal Court. *International Criminal Law Review, 17*(4), 625–655.

Doob, A. N., & Webster, C. M. (2003). Sentence Severity and Crime. *Crime and Justice, 30*, 143–195.

Drumbl, M. (1998). Waging War Against the World. *Fordham International Law Journal, 22*(1), 122–153. https://ir.lawnet.fordham.edu/ilj/vol22/iss1/3
Duff, R. A. (2001). *Punishment, Communication, and Community*. Oxford University Press.
Du Rées, H. (2001). Can Criminal Law Protect the Environment? *Journal of Scandinavian Studies in Criminology and Crime Prevention, 2*(2), 109–126. https://doi.org/10.1080/140438501753737606
End Ecocide in Europe. (2012). A Citizens' Initiative to Give the Earth Rights. https://www.europarl.europa.eu/meetdocs/2014_2019/documents/peti/dv/eciendecocideineurope_/eciendecocideineurope_en.pdf
End Ecocide on Earth. (2016, September). Ecocide Amendments Proposal. https://www.endecocide.org/wp-content/uploads/2016/10/ICC-Amendements-Ecocide-ENG-Sept-2016.pdf
End Ecocide on Earth. (2024). Comment for the Consultation from the OTP of the ICC on the Use of the Existing Crimes in the Rome Statute to Prosecute Environmental Crimes. https://www.endecocide.org/wp-content/uploads/2024/03/OTP-consultation-End-Ecocide-on-Earth-response.pdf
European Parliament, Committee on the Environment, Public Health and Food Safety. (2022). Opinion (25 October 2022) for the Committee on Legal Affairs on the Proposal for a Directive of the European Parliament and of the Council on the Protection of the Environment through Criminal Law and Replacing Directive 2008/99/EC, C9-0466/2021.
Falk, R. A. (1973). Environmental Warfare and Ecocide-Facts, Appraisal, and Proposals. *Revue Belge de Droit International, 9*(1), 1–27.
Floyd, R. (2015). Environmental Security and the Case Against Rethinking Criminology as Security-ology. *Criminology and Criminal Justice, 15*(3), 277–282. https://doi.org/10.1177/1748895815584720
Fried, J. (1973). War by Ecocide: Some Legal Observations. *Paper in the Effects of Modern Weapons on the Human Environment in Indochina*, June 1972.
Gauger, A., Rabatel-Fernel, M., Kulbicki, L., Short, D., & Higgins, P. (2012). Ecocide is the Missing 5th Crime Against Peace. The Ecocide Project, Human Rights Consortium.
Glasgow, J. (2015). The Expressivist Theory of Punishment Defended. *Law and Philosophy, 34*(6), 601–631. https://www.jstor.org/stable/24572388
Gray, M. (1996). The International Crime of Ecocide. *California Western International Law Journal, 26*(2), 215–272.
Hansen, T. O. (2025). Towards a New Agenda for Ecocide Norms—Adoption and Enforcement. In R. Pereira, & T. Fajardo (Eds.), *A Research Agenda for Environmental Crime and the Law* (pp. 65–90). Edward Elgar.

Haines, F., & Parker, C. (2018). Moving Towards Ecological Regulation: The Role of Criminalisation. In C. Holley, & C. Shearing (Eds.), *Criminology and the Anthropocene* (pp. 81–102). Routledge.

Hart, H. M. (1958). The Aims of the Criminal Law. *Law and Contemporary Problems, 23*(3), 401–441.

Higgins, P. (2012). *Eradicating Ecocide*. Shepheard-Walwyn.

Higgins, P., Short, D., & South, N. (2013). Protecting the Planet. *Crime, Law and Social Change, 59*(3), 251–266. https://doi.org/10.1007/s10611-013-9413-6

Hillyard, P., & Tombs, S. (2007). From Crime to Social Harm? *Crime, Law and Social Change, 48*(1), 9–25. https://doi.org/10.1007/s10611-007-9079-z

Hodgson, N. (2021). Exploring the International Criminal Court's Deterrent Potential. *Journal of International Criminal Justice, 19*(4), 913–938.

Hodgson, N. (2024). Resisting the State Crimes of the Global North: Exploring the Potential of International Criminal Law. *Journal of International Criminal Justice, 22*(1), 151–168. https://doi.org/10.1093/jicj/mqae010

ICC. (1998). *Rome Statute of the International Criminal Court,* in force on 1 July 2002, UNTS, vol. 2187, No. 38544.

Independent Expert Panel (2021, June). *Commentary and Core Text.* https://www.stopecocide.earth/legal-definition

Jo, H., & Simmons, B. A. (2016). Can the International Criminal Court Deter Atrocity? *International Organization, 70*(3), 443–475. https://doi.org/10.1017/S0020818316000114

Jowa, E. (2024, June 15). Princess Esmeralda: "Only the Recognition of Ecocide as an International Crime will Prevent Damage to the Climate." *Paris Match.* https://www.parismatch.be/actualites/environnement/2024/06/15/princesse-esmeralda-seule-la-reconnaissance-de-lecocide-en-tant-que-crime-international-permettra-de-prevenir-les-dommages-a-lencontre-du-climat-BW6ZG7DTUVHWBJ7BBGRWHVKFVU/

Kant, I. (1838). *Die Metaphysik der Sitten.* Modes und Baumann.

Killean, R. (2025). Ecocide's Evolving Relationship with War. *Environment and Security*. https://doi.org/10.1177/27538796251347111

Killean, R., & Short, D. (2023). 'Scoping a Domestic Crime of Ecocide' Report for the Environmental Rights Centre for Scotland.

Killean, R., & Short, D. (2025, April). A Critical Review of the Law and Policy of Ecocide. *Environmental Politics*, 1–21.

Knight, B. (2020, July 16). Do Harsher Punishments Deter Crime? *UNSW Newsroom.* https://www.unsw.edu.au/newsroom/news/2020/07/do-harsher-punishments-deter-crime

Knox, R., & Whyte, D. (2025). Law's Quick Fix? Ecocide, Social Transformation and the Pitfalls of Criminalisation. *Environmental Politics*, 1–20. https://doi.org/10.1080/09644016.2025.2492443

Koop, F. (2024, January 16). Ecocide: What is it and Should it be Considered an International Crime? *Dialogue Earth.* https://dialogue.earth/en/nature/387886-ecocide-what-is-it-and-should-it-be-considered-an-international-crime/

Lennon, M. (2023, November 8). *A Proposal for a Bill to Protect the Environment in Scotland and Deter Environmental Damage by Introducing the Crime of Ecocide into Scots Law.* Consultation.

Lennon, M., & Schwartz, J. (2024, August 19). The Promise of Ecocide Law, from Boardroom to Courtroom (Commentary). *Mongabay.* https://news.mongabay.com/2024/08/the-promise-of-ecocide-law-from-boardroom-to-courtroom-commentary/

Lynch, M. J., Burns, R. G., & Stretesky, P. (2014). *Environmental Law, Crime, and Justice* (2nd ed.). LFB Scholarly.

Lynch, M. J., Fegadel, A., & Long, M. A. (2020). Green Criminology and State-Corporate Crime. *Journal of Genocide Research, 23*(2), 236–256. https://doi.org/10.1080/14623528.2020.1771998

Lynch, M. J., Long, M. A., Barrett, K. L., & Stretesky, P. B. (2013). Is It a Crime to Produce Ecological Disorganization? *British Journal of Criminology, 53*(6), 997–1016. https://doi.org/10.1093/bjc/azt051

Maddrick, A. (2024, February 11). Our Common Heritage: The Role of Ecocide Law. *Eco-Connect.* https://eco-nnect.com/our-common-heritage-the-role-of-ecocide-law/

Mavronicola, N. (2024). The Case Against Human Rights Penality. *Oxford Journal of Legal Studies, 44*(3), 535–562.

McAuliffe, P. (2012). Suspended Disbelief? The Curious Endurance of the Deterrence Rationale in International Criminal Law. *New Zealand Journal of Public and International Law, 10*(2), 227–262.

McEvoy, K. (2007). Beyond Legalism. *Journal of Law & Society, 34*(4), 411–440. https://doi.org/10.1111/j.1467-6478.2007.00399.x

McKie, R. E., Stretesky, P. B., Lynch, M. J., & Long, M. A. (2017). Green Criminology and the Prevention of Ecological Destruction. In T. Bergin, & E. Orlando (Eds.), *Forging a Socio-Legal Approach to Environmental Harms* (pp. 20–39). Routledge.

Mégret, F. (2021). The Anti-Deterrence Hypothesis. *Journal of International Criminal Justice, 19*(4), 859–876. https://doi.org/10.1093/jicj/mqaa015

Mehta, J. (2019, April 26). Polly Higgins Obituary. *The Guardian.* https://www.theguardian.com/environment/2019/apr/25/polly-higgins-obituary

Mehta, J., & Jackson, J. (2021, February 24). To Stop Climate Disaster, Make Ecocide an International Crime. It's the Only Way. *The Guardian.* https://www.theguardian.com/commentisfree/2021/feb/24/climate-crisis-ecocide-international-crime

Minkova, L. G. (2023). The Fifth International Crime: Reflections on the Definition of "Ecocide." *Journal of Genocide Research, 25*(1), 62–83. https://doi.org/10.1080/14623528.2021.1964688

Moore, M. S. (2010). *Placing Blame: A Theory of the Criminal Law.* Oxford University Press.

Netherlands Consultation. (2024, February 1). Chapter 7 Consultation report. https://www.internetconsultatie.nl/ecocidewet/berichten

Nixon, R. (2011). *Slow Violence and the Environmentalism of the Poor.* Harvard University Press.

Ogus, A., & Abbott, C. (2002). Sanctions for Pollution: Do We Have the Right Regime? *Journal of Environmental Law, 14*(3), 283–298. https://www.jstor.org/stable/44248387

Orlando, E., & Bergin, T. (2017). *Forging a Socio-Legal Approach to Environmental Harms: Global Perspectives.* Routledge.

Ragnarsdóttir, K. V., & Sharpston, E. (2023, December 20). Scientists, Lawyers Unite Behind Ecocide Law. *Ecologist.* https://theecologist.org/2023/dec/20/scientists-lawyers-unite-behind-ecocide-law#:~:text=Ecocide%20law%20gives%20us%20the,International%20Criminal%20Court%20(ICC)

Robinson, D. (2022). Ecocide—Puzzles and Possibilities. *Journal of International Criminal Justice, 20*(2), 313–347. https://doi.org/10.1093/jicj/mqac021

Ruhashyankiko, N. (1978, July 4). *United Nations Sub-Commission on Prevention of Discrimination and Protection of Minorities. Study of the Question of the Prevention and Punishment of the Crime of Genocide. Prepared by Mr. Nicodème Ruhashyankiko* (E/CN.4/Sub.2/416).

Sarre, R. (2024). *Preventing Crime.* Springer.

Schell-Busey, N., Simpson, S. S., Rorie, M., & Alper, M. (2016). What Works? *Criminology & Public Policy, 15*(2), 387–416. https://doi.org/10.1111/1745-9133.12195

Schnyder, M. (2023). Norm-Building in Advocacy to Establish Ecocide as an International Crime. *Social Alternatives, 42*(4), 55–61. https://search.informit.org/doi/10.3316/informit.T2024110800012391951250748

Schwendinger, H., & Schwendinger, J. (1977). Social Class and the Definition of Crime. *Crime and Social Justice, 7*, 4–13.

South, N. (2017). Environmental Crimes and Harms: A Green Criminology Approach and Socio-Legal Challenges. In E. Orlando, & T. Bergin (Eds.), *Forging a Socio-Legal Approach to Environmental Harms: Global Perspectives* (pp. 20–38). Routledge.

Stop Ecocide Italia. (2024). *Response to Monica Lennon Proposed Ecocide (Prevention) (Scotland) Bill*. www.ecocidelaw.scot

Tomuschat, C. (1996). Crimes Against the Environment. *Environmental Policy and Law, 26*(6), 242–243.

Van Raan, L. (2023). *Proposal of Law by Member Van Raan Containing Amendment of the Criminal Code and the Code of Criminal Law in Connection with the Criminalization of Ecocide (Ecocide Criminalization Act), Explanatory Memorandum.*

Wellsmith, M. (2011). Wildlife Crime: The Problems of Enforcement. *European Journal on Criminal Policy and Research, 17*(2), 125–148. https://doi.org/10.1007/s10610-011-9140-4

Westerhuis, D., Walters, R., & Wyatt, T. (2013). *Emerging Issues in Green Criminology: Exploring Power, Justice and Harm*. Palgrave Macmillan.

Whitaker, B. (1985). *Revised and Updated Report on the Question of the Prevention and Punishment of the Crime of Genocide, UN ESCOR, Human Rights Sub-Commission on the Prevention of Discrimination and Protection of Minorities, 38th Sess.*, UN Doc. E/CN.4/Sub.2/1985/6.

White, R. (2008). The Criminalisation of Environmental Harm. *Criminal Justice Matters, 74*(1), 35–37. https://doi.org/10.1080/09627250802476817

White, R. (2011). Environmental Law Enforcement. *Australasian Policing, 3*(1), 12–16.

White, R. (2018). Ecocide and the Carbon Crimes of the Powerful. *University of Tasmania Law Review, 37*, 95–114.

White, R. (2024). Environmental Harms and Innovative Justice. In S. Smith, & I. Sahramäki (Eds.), *Research Handbook on Environmental Crimes and Criminal Enforcement* (pp. 128–145). Edward Elgar.

White, R., & Kramer, R. (2015). Critical Criminology and the Struggle Against Climate Change Ecocide. *Critical Criminology, 23*(4), 383–399. https://doi.org/10.1007/s10612-015-9292-5

Whyte, D. (2016). *Ecocide*. Manchester University Press.
Zierler, D. (2011). *The Invention of Ecocide: Agent Orange, Vietnam, and the Scientists Who Changed the Way We Think About the Environment*. University of Georgia Press.

12

A Kaleidoscopic Justice for Green Criminology

Rachel Killean, Kajsa Lundberg, and Mark Hamilton

1 Introduction

At the outset of this book, we posed a central question: what does justice mean in the face of intersecting ecological crises and the widening scope of victims, human and other-than-human alike? We highlighted the proliferation of justice frameworks within green criminology (e.g. environmental, ecological, species, climate, interspecies) and the urgent need to think through their overlaps and tensions. This volume has brought

R. Killean (✉)
University of Sydney, Sydney, NSW, Australia
e-mail: rachel.killean@sydney.edu.au

K. Lundberg
Centre for Urban Research, School of Global, Urban and Social Studies, RMIT University, Melbourne, VIC, Australia
e-mail: kajsa.lundberg@rmit.edu.au

M. Hamilton
Australian Catholic University, Sydney, NSW, Australia
e-mail: mark.hamilton@acu.edu.au

R. Killean et al. (eds.), *Conceptualising Justice*, Palgrave Studies in Green Criminology, https://doi.org/10.1007/978-3-032-19791-7_12

together a range of perspectives on what justice means in the context of environmental harm and other-than-human victimisation. Located in the Antipodean context and shaped by contributions from Australia and Aotearoa/New Zealand, this collection has been grounded in the lived realities of bushfires, biodiversity loss, and settler-colonial water governance, while also engaging global debates around ecocide and ecological justice.

In this final chapter, we aim to highlight the common themes and connections present across the diverse approaches to 'green' justice explored in this anthology. These include visions of justice that encompass recognition, participation (both human and non-human), and relationality, extending beyond traditional concepts of economic, social, or environmental justice. As such, this chapter brings these contributions together to draw some overarching conclusions regarding current conceptualisations of justice within the field of green criminology. To do so, we take inspiration from Clare McGlynn and Nicole Westmarland's (2019) framework of kaleidoscopic justice. Developed in response to a series of interviews with victim-survivors of sexual violence, a kaleidoscopic justice approach rejects the assumption that justice is linear, delivered solely by the state, and predominantly retributive. Instead, justice is better understood as a continually shifting pattern: lived, ongoing, collective, and responsive to those experiencing harm. When applied to green criminology, kaleidoscopic justice allows us to hold together the many overlapping frameworks presented in this book, and to acknowledge the dynamic, plural, and relational ways in which justice might be imagined for environmental destruction and related suffering. In doing so, we return to the book's starting point: the call to rethink justice in the Anthropocene, beyond the limits of retribution and the human-centred state.

One theme that emerged throughout the book was that of other-than-human victimhood, agency and claims to justice. To unpack this theme's relevance to kaleidoscopic justice, we draw from new materialism and Science and Technology Studies. We highlight the importance of shifting attention from anthropocentric conceptions of agency as something inherently human (with related justice implications) to consider the ways humans, other-than-human species, ecologies, and environments co-constitute the world and hierarchies within it. Reflecting the themes explored throughout the book, we sketch out an overlapping and interlinked approach that prioritises meaningful consequences, inclusivity, participation (of humans and other-than-humans), and—importantly—relationality.

2 Towards Kaleidoscopic Justice

In their work with sexual violence victim-survivors, Clare McGlynn and Nicole Westmarland (2019: 185) have drawn attention to the 'myriad perceptions, ideas and suggestions which might, in some shape or combination, provide—for some—a sense of justice'. They term this approach to justice 'kaleidoscopic justice', evoking the kaleidoscope's ability to produce an ever-changing viewed pattern, 'constantly refracted through new circumstances and understandings' (McGlynn & Westmarland, 2019: 186). For their research participants, justice was not reducible to a single outcome such as conviction or punishment. Rather, it was experienced through a constellation of interlinked elements: recognition, voice, participation, dignity, and consequences. Each of these dimensions could shift in salience depending on the context, and each mattered in terms of shaping how victim-survivors understood justice as lived and ongoing. McGlynn and Westmarland's resulting framework emphasises plurality and process; it challenges linear, state-centric accounts of justice and instead insists on a more dynamic, inclusive, and relational orientation.

Although developed in the context of sexual violence, we argue that the kaleidoscopic justice framework provides a powerful lens for thinking about justice more broadly. Its central insights are not necessarily bound to the dynamics of sexual harm but instead speak to how justice is experienced in contexts where conventional, state-centric, and retributive frameworks prove profoundly insufficient. Indeed, ecofeminists, Indigenous and decolonial feminists have long interrogated the connection between sexual and environmental harms and have argued for a more expansive justice for many years (Cabnal, 2010; Plumwood, 1999, 2002; Vargas & Killean, 2023). In the context of green criminology, the harms under examination may encompass some that are different in nature, but that are likewise diffuse, multi-layered, and not easily captured by traditional remedies. Kaleidoscopic justice offers a language and framework capable of accommodating that complexity, helping us draw together the diverse ways justice is conceptualised across the chapters of this volume.

Kaleidoscopic justice offers an alternative, more inclusive way of thinking about justice than that offered by the criminal justice system alone. As such, it is a framework that is sensitive to both the *value and the limitation of criminal justice* mechanisms and outcomes. This sensitivity resonates with the recurring ambivalence towards law and criminalisation within this book. On the one hand, several chapters highlight the promise of legal recognition and criminal sanction. The Australian Capital Territory's (ACT) recognition of animal sentience (McKinley, Vachette & Honey, Chapter 7), the criminal prosecutions of environmental offences in New South Wales (NSW) (Wright, Chapter 5), the revisions of Vietnam's penal code to address wildlife trafficking (Smith, Chapter 10), restorative justice interventions following environmental prosecutions (Hamilton, Chapter 6) and the ongoing international campaigns to criminalise ecocide all represent significant steps in affirming responsibility for environmental harm (Killean, Chapter 11). On the other hand, these same chapters caution against over-reliance on criminalisation as a means of delivering justice. For example, Rachel Killean notes that the campaign to add ecocide to the Rome Statute, while powerful in its symbolic, preventative, and expressive aims, may promise more than international criminal law can deliver in practice. The deterrent potential of international crimes remains uncertain, and political realities often constrain outcomes. Michael Smith shows how deterrence logics risk re-inscribing anthropocentric framings of justice, casting rhino protection in Vietnam as a matter of human security rather than recognising rhinos as victims in their own right. Sarah Wright observes the tensions that arise when public participation disrupts prosecutorial discretion. Mark Hamilton stresses that criminalisation cannot reach the vast areas of lawful but harmful activities that cause environmental harm. These analyses reveal that law may offer recognition, symbolism, and deterrence, but it presents—at best—just one fragment in the kaleidoscope.

A concern over the limits of criminalisation should not be conflated with a disinclination for *consequences*. Indeed, as Smith's chapter notes, punishment remains a key focus for many environmental campaigners (Chapter 10). Rather, our authors highlight the need for consequences to be conceptualised in more holistic ways, looking beyond a traditional focus on retribution to grapple with the needs and wishes of victims and broader society. Examples of how this might be done are highlighted throughout the book. For example, Hamilton understands consequences as being meaningful when they restore what has been damaged, requiring those harmed to retain or regain their capability to function (Chapter 6). Similarly, Sarah Wright argues that sentencing for environmental prosecutions can deliver more than fines or imprisonment, with orders requiring offenders to fund rehabilitation projects or environmental enhancements arguably offering a more meaningful response to the harm caused (Chapter 5). Several authors argue for consequences that move beyond repair to encompass more transformative goals. For example, Cassie Pedersen and Stephen Burrell stress that justice must involve shifts in structural arrangements that alter the patriarchal and colonial logics that underlie ecological harm (Chapter 4). Similarly, Alexander Baird and Reece Walters argue for redistributive justice in response to water injustices, comprising a rethinking of water governance that honours Indigenous knowledge and repairs historical inequities (Chapter 3). For Kajsa Lundberg, meaningful consequences would require reform to urban planning, reducing vulnerability rather than simply sanctioning those responsible (Chapter 8). Such arguments look beyond individual incidents of harm and response, bringing together backward-looking justice elements with more future-oriented, preventative conceptualisations of social justice.

Alongside meaningful consequences, McGlynn and Westmarland emphasise the importance of *recognition* as a core aspect of kaleidoscopic justice. They argue that the standard outcomes that can be expected from the criminal justice system will do little to address moral injuries unless those outcomes are accompanied by measures of recognition. Recognition, as they conceptualise it, is more than being believed. It encompasses an expressive dimension—the 'significance of the experience being acknowledged; of its power and importance for the victim-survivor and in society more generally' (McGlynn & Westmarland, 2019: 188). Drawing on Nancy Fraser's argument that recognition has an important collective element (1998; see also 2003), they highlight the injustice caused by misrecognition and argue for forms of justice that effect broader social change and foster a sense of connectedness (McGlynn & Westmarland, 2019: 189, 194).

This book's chapters consistently highlight recognition as a core aspect of justice, whether or not they engage with the concept explicitly. Recognition of animal sentience in law (McKinley, Vachette & Honey, Chapter 7), proposals to extend welfare protections to invertebrates (Durrant, Chapter 9), the naming of ecocide as a crime (Killean, Chapter 11), and the capacity of communities to appear in court on behalf of ecosystems (Wright, Chapter 5)—all of these acts matter not only for their material impacts or consequences, but for the cultural work they perform. They reframe societal understandings of who and what counts as a victim, and they affirm the dignity and visibility of those previously excluded from the justice imaginary. Moreover, approaches such as ecofeminist forms of 'just care', as advocated by Pedersen and Burrell (Chapter 4), recognise not only marginalised victims but marginalised forms of knowledge, such as those held by Indigenous peoples. This recognition is arguably particularly crucial in the context of green criminology. As mentioned in several chapters, the recognition of alternative knowledge systems often facilitates a more inclusive understanding of victimhood and agency (Baird & Walters, Chapter 3; Lundberg, Chapter 8; Pedersen and Burrell, Chapter 4). More specifically, such knowledge systems point to the *victimhood and agency of other-than-human entities* (for example, Durrant, Chapter 9). We continue this theme in the following section, which takes a closer look at how justice might be expanded beyond the human, and the importance of participation and the recognition of relational agency in such an expansion.

3 Expanding Justice Beyond the Human

Traditional criminological conceptions of justice tend to be characterised by both anthropocentrism and individualism. This is especially true for conceptions of justice associated with the criminal justice system, which tends to focus on natural persons as both victims and perpetrators (Whyte, 2020). Agency is central to this understanding of justice, underpinning our approach to—for example—perpetrators' competency to stand trial, the actions of victims in the context of assessing 'ideal victimhood' (Christie, 1986), and crime prevention strategies grounded in rational choice perspectives. Most justice perspectives explored in this book, however, fall outside such neat, individual and distinctively human conceptions of justice, often invoking state-corporate perpetrators and human and non-human victims of harm. This requires alternative ways of understanding not only justice but the agentic relations involved in experiences of (in)justice.

Throughout the book, authors consistently resist narrow or anthropocentric framings of justice. Rob White's opening chapter provides the conceptual foundation for such an approach by developing the eco-justice framework, which integrates environmental justice (focused on humans), ecological justice (focused on ecosystems), and species justice (focused on animals and plants), with the additional dimension of extraterrestrial justice (Chapter 2). This expansive orientation set the stage for subsequent contributions, all of which seek to move beyond anthropocentrism. For example, Pedersen and Burrell explicitly interrogate the legacies of Enlightenment thought that have shaped dominant conceptions of justice. They argue that ecofeminist theory, with its emphasis on relationality and care, offers a way to resist masculinised, anthropocentric frameworks and to instead conceive justice as 'just care', including for the more-than-human world (Chapter 4). Pedersen and Burrell use salmon to illustrate how criminologists might push for justice for species 'markedly different to humans', while Russil Durrant's chapter pushes even further. His chapter brings green criminology into novel territory by advocating for invertebrate justice, insisting that the sheer scale of harms inflicted on insects and other invertebrates, coupled with their ecological importance, makes them a vital site of criminological concern (Chapter 9). Amber McKinley, Astrid Vachette, and Erica Honey bring this expansion into the legal realm by analysing the ACT's explicit recognition of animal sentience (Chapter 7). Their chapter illustrates how this reform unsettles property-based legal categories, while also revealing the persistence of industry exemptions and anthropocentric limitations.

Together, the book's chapters each underline a consistent theme: conceptualisations of 'justice' in green criminology cannot be confined to humans. Instead, they must account for animals, ecosystems, and neglected life forms, demanding a more inclusive, relational, and multi-species approach. These examples make clear that there are complex relationships between human and other-than-human agency that constitute social phenomena and, as such, justice. To understand these complex entanglements and the assemblages of actors and agency that constitute in(justice), we suggest that the field of green criminology might draw on the insights from Science and Technology Studies, an area of growing recognition across the social sciences (see also Natali et al., 2023).

In line with kaleidoscopic justice, Science and Technology Studies approaches emphasise plurality and process, challenging traditional, linear understandings of socio-natural phenomena. Moreover, Science and Technology Studies conceptualise agency as the assembly of diverse entities, striving to transcend binary distinctions between 'nature' and 'society' and fostering an integration of natural and social sciences (Latour, 2000). Science and Technology Studies, therefore, draw attention to the relations between humans and other-than-human entities, interrogating how these relationships constitute the socio-natural world (Latour, 2000). According to this approach, agency is understood as created relationally in the interactions between humans and other-than-human actors. This is particularly true for Actor-Network Theory, which seeks to understand agency through the concept of networks (O'Lear, 2016), drawing attention to 'ignored or under-researched relations, especially between humans and non-humans' (Schumacher, 2023: 255–256).

As noted by Schumacher (2023), the strengths of Science and Technology Studies and Actor-Network Theory lie in part in their ability to be combined with other perspectives. Applying this thinking to justice in green criminology provides a lens through which to recognise and engage with agency. It encourages an approach that frames agency as the result of a combination of human and other-than-human actors that can prevent, construct and suffer from environmental harms, and must be considered in any conception of justice. This approach is foregrounded throughout this book, which rarely treats agency as a fixed attribute of rational, autonomous humans. Instead, agency is consistently problematised, expanded, and relationalised.

Bringing this back to our adopted framework of kaleidoscopic justice, a framing of relational agency has particular resonance with the reimagining of justice as participatory and processual, a core aspect of McGlynn and Westmarland's framework and a recurrent theme in this book. For example, Hamilton develops justice as meaningful involvement through restorative justice conferencing, an approach which encompasses recognition of human and other-than-human victims and the participation of 'a wide range of people, operators, communities, the environment (and its constituent parts) and countries' (Chapter 6). Wright's chapter explores the more limited possibilities of participation in formal criminal justice processes, nonetheless highlighting pathways through which the public and community groups might 'speak for the environment' (Chapter 5). Focusing on more future-oriented justice processes, Baird and Walters' chapter on 'water justice' argues for a 'more grounded, relational approach' to water governance, which foregrounds the voices of traditionally marginalised communities, but also recognises freshwater bodies as 'independent beings with agency' (Chapter 3). Across these chapters, justice emerges as a process through which the relationships between humans and other-than-human entities find (albeit in some cases rather limited) expression. In McGlynn and Westmarland's terms, these examples illustrate the kaleidoscopic dimensions of participation, reminding us that justice is as much about being recognised, treated with dignity, and included in decision-making processes as it is about outcomes.

Bringing green criminology and Science and Technology Studies together also offers opportunities to reflect on how structural factors shape the exercise of agency. As Castree (2002) outlines, applying critical and green theoretical approaches to Actor-Network Theory can deepen its understanding of how power shapes networks. Such an approach maintains Actor-Network Theory's critique of binaries and recognition of a more expansive agency, while at the same time recognising that there are external processes that structure the ways that agency is enacted and socio-natural networks play out in the world (Castree, 2002: 135). The external constraints placed on agency emerge across many chapters in this book. Pedersen and Burrell highlight the entanglement of patriarchy, colonialism, and capitalism in ecological destruction, positioning ecofeminism as a means of exposing these interlinked logics (Chapter 4). Lundberg's chapter highlights how agency is shaped by urban planning, with disasters revealing how socio-spatial arrangements enable or disable the ability of communities and environments to respond (Chapter 8). Killean's chapter notes that the pursuit of justice for the human and other-than-human victims of ecocide is likely to be constrained and shaped by geopolitical and corporate interests (Chapter 11). Similarly, McKinley, Vachette and Honey illustrate how industry exemptions and corporate interests continue to shape legal responses to harms against animals (Chapter 7). These chapters reinforce the need to challenge structural drivers of environmental destruction. Future research that continues to bring Science and Technology Studies into dialogue with green criminology may further reveal the possibilities and limitations of meaningful engagement with human and other-than-human agency.

Bringing these diverse strands of justice together, what emerges is a vision of justice that resonates closely with McGlynn and Westmarland's kaleidoscopic framework, while also illustrating the framework's adaptability to other forms of harm and victimised entities. Just as a kaleidoscope produces ever-shifting patterns from multiple fragments of coloured glass, the section and the preceding section demonstrate how justice for environmental harm can be conceptualised through shifting and overlapping recognitions of human, animal, ecological, and even invertebrate agency. Each turn of perspective, whether that be eco-justice, ecofeminist care, species sentience, invertebrate protection, or deterrence, creates a different but equally vital configuration of justice. The kaleidoscopic lens enables us to hold these multiple, sometimes competing, accounts together without reducing them to a single hierarchy of value.

In this way, expanding justice beyond the human is not an isolated endeavour but part of a broader, adaptive pattern: a justice that is relational, inclusive, and continually reconfigured in response to the lived realities of harm across species and ecosystems.

4 Green Kaleidoscopic Justice

Taken together, the chapters in this book illustrate what kaleidoscopic justice looks like when refracted through green criminology. Moving between environmental, ecological, species, invertebrate and even extraterrestrial justice; between deterrence, care, reparation, participation, and symbolism; between law, activism, planning, and culture, a green kaleidoscopic justice does not promise neat solutions or singular frameworks but insists on justice as an evolving, relational, and multidimensional practice.

Insights from Science and Technology Studies and Actor-Network Theory enrich the kaleidoscopic metaphor. Justice emerges not only through plural perspectives but through assemblages of human and other-than-human actors, whose interactions co-constitute both harm and response. As Latour (2000) reminds us, agency is not an individual property but a relational effect of networks; as such, justice too must be understood as something enacted through these socio-natural entanglements. The chapters in this book illustrate this: rivers and freshwater bodies claim recognition in water governance, urban infrastructures amplify or constrain vulnerability, animals and invertebrates are repositioned as victims, and legal reforms are shaped by the persistent influence of corporate and geopolitical power.

A kaleidoscopic justice informed by Science and Technology Studies thus draws our attention to the networks through which justice is made and unmade. Each turn of the kaleidoscope does not simply reveal different normative visions of justice, but also different configurations of actors, agencies, and infrastructures: a courtroom in New South Wales where communities seek to speak for koalas, a legislative chamber in the ACT recognising animal sentience amid industry exemptions, a planning inquiry in Melbourne where urban design structures vulnerability, or an international campaign for ecocide where ecosystems, activists, states, and corporations collide.

By embracing this plurality and material relationality, green criminology can better respond to the complexities of ecological harm. A kaleidoscopic justice informed by Science and Technology Studies and Actor-Network Theory asks us to see justice as not only plural, inclusive, and evolving, but also as enacted through networks of humans, animals, technologies, and ecologies. Such an approach better equips us to imagine responses to environmental harm that are both adaptive and transformative in the Anthropocene.

Finally, it is important to return to where this book began. In our introduction, we highlighted the Antipodean context. For our authors, theorising green criminological approaches to justice occurred against the backdrop of a region marked by devastating bushfires, floods, biodiversity loss, and the ongoing legacies of settler-colonial environmental governance. The kaleidoscopic justice lens allows us to hold these harms and visions of justice together, to treat them as fragments of a broader vision that is continually shifting, relational, and alive to both local and global struggles. Just as the kaleidoscope refracts light into ever-changing patterns, so too does justice in the Anthropocene emerge through new constellations of recognition, participation, and consequence. This, we suggest, may prove to be the enduring contribution of green criminology: to keep turning the kaleidoscope so that justice remains adaptive, inclusive, and responsive to the urgent demands of a world in crisis.

References

Cabnal, L. (2010). *Feminismos Diversos: El Feminismo Comunitario*. ACSUR-Las Segovias.

Castree, N. (2002). False Antitheses? Marxism, Nature and Actor-Networks. *Antipode, 34*(1), 111–146. https://doi.org/10.1111/1467-8330.00228

Christie, N. (1986). The Ideal Victim. In E. Fattah (Ed.), *From Crime Policy to Victim Policy* (pp. 17–30). Palgrave Macmillan.

Fraser, N. (1998). *Social Justice in the Age of Identity Politics: Redistribution, Recognition, Participation* (Discussion Paper).

Fraser, N. (2003). *Recognition Struggles and Social Movements: Contested Identities, Agency and Power*. Cambridge University Press.

Latour, B. (2000). When Things Strike Back: A Possible Contribution of 'Science Studies' to the Social Sciences. *The British Journal of Sociology, 51*(1), 107–123. https://doi.org/10.1111/j.1468-4446.2000.00107.x

McGlynn, C., & Westmarland, N. (2019). Kaleidoscopic Justice: Sexual Violence and Victim-Survivors' Perceptions of Justice. *Social & Legal Studies, 28*(2), 179–201. https://doi.org/10.1177/0964663918761200

Natali, L., Suman, A. B., & de Nardin Budó, M. (2023). What About Environmental 'Victims'? Methodological Reflections for an Activist Criminology. In V. Canning, G. Martin, & S. Tombs (Eds.), *The Emerald International Handbook of Activist Criminology* (pp. 215–231). Emerald Publishing Limited.

O'Lear, S. (2016). Climate Science and Slow Violence: A View From Political Geography and STS on Mobilizing Technoscientific Ontologies of Climate Change. *Political Geography, 52*, 4–13. https://doi.org/10.1016/j.polgeo.2015.01.004

Plumwood, V. (1999). Ecological Ethics from Rights to Recognition: Multiple Spheres of Justice for Humans, Animals and Nature. In N. Low (Ed.), *Global Ethics and Environment* (pp. 188–212). Routledge.

Plumwood, V. (2002). *Feminism and the Mastery of Nature*. Taylor and Francis.

Schumacher, J. M. (2023). Framing REDD+: Political Ecology, Actor–Network Theory (ANT), and the Making of Forest Carbon Markets. *Geographica Helvetica, 78*(2), 255–265. https://doi.org/10.5194/gh-78-255-2023

Vargas, D. S., & Killean, R. (2023). Women's Experiences of Environmental Harm in Colombia: Learning from Black, Decolonial and Indigenous Communitarian Feminisms. In E. Milne, P. Davies, J. Heydon, K. Peggs, & T. Wyatt (Eds.), *Gendering Green Criminology* (pp. 229–250). Bristol University Press.

Whyte, D. (2020). *Ecocide: Kill the Corporation Before it Kills Us*. Manchester University Press.

Index

R. Killean et al. (eds.), *Conceptualising Justice*, Palgrave Studies in Green Criminology, https://doi.org/10.1007/978-3-032-19791-7

The manufacturer's authorised representative in the EU is Springer Nature Customer Service Centre GmbH, Europaplatz 3, 69115 Heidelberg, Germany. If you have any concerns regarding our products, please contact ProductSafety@springernature.com

Printed and bound by CPI Group (UK) Ltd, Croydon, CR0 4YY

17/09/2026

02218433-0005